# Recycling Strategy and Challenges Associated with Waste Management Towards Sustaining the World

*Edited by Hosam M. Saleh and Amal I. Hassan*

Published in London, United Kingdom

Recycling Strategy and Challenges Associated with Waste Management Towards Sustaining the World
http://dx.doi.org/10.5772/intechopen.104051
Edited by Hosam M. Saleh and Amal I. Hassan

Contributors
Hesham Ali, Aditia Rojali, Bukola Margaret Popoola, Michael Naor, Hiromichi Fumoto, María Noel Cabrera, Leonardo Clavijo, Mairan Guigou, Norberto Cassella, Claudia Lareo, Mario Daniel Ferrari, Florencia Risso, Gastón Cortizo, Lucía Velazco, Kachikoti Banda, Erastus M. Mwanaumo, Bupe Getrude Mwanza, Elisabeth Eppinger, Alina Slomkowski, Tanita Behrendt, Sigrid Rotzler, Max Marwede, Patchiya Phanthong, Shigeru Yao, Florence Akinyi Ogutu, Bessy Kathambi



First published in London, United Kingdom, 2023 by IntechOpen
IntechOpen is the global imprint of INTECHOPEN LIMITED, registered in England and Wales, registration number: 11086078, 5 Princes Gate Court, London, SW7 2QJ, United Kingdom

British Library Cataloguing-in-Publication Data
A catalogue record for this book is available from the British Library

Additional hard and PDF copies can be obtained from orders@intechopen.com

Recycling Strategy and Challenges Associated with Waste Management Towards Sustaining the World
Edited by Hosam M. Saleh and Amal I. Hassan
p. cm.
Print ISBN 978-1-83768-011-5
Online ISBN 978-1-83768-012-2
eBook (PDF) ISBN 978-1-83768-013-9

For EU product safety concerns:
IN TECH d.o.o., Prolaz Marije Krucifikse Kozulić 3, 51000 Rijeka, Croatia,
info@intechopen.com or visit our website at intechopen.com.

# Meet the editors

Hosam M. Saleh is a Professor of Radioactive Waste Management at the Radioisotope Department, Atomic Energy Authority, Egypt. He obtained an MSc and Ph.D. in Physical Chemistry from Cairo University, Egypt. He has more than 25 years of experience in hazardous waste management with an emphasis on the treatment and development of new matrixes for the immobilization of these wastes. He is also interested in studying innovative economic and environmentally friendly techniques for the management of hazardous and radioactive wastes. Dr. Saleh has authored many peer-reviewed scientific papers and chapters and edited several books. He has been listed among the top 2% of scientists in the world by Stanford University in 2020, 2021, and 2022.

Amal I. Hassan is a Professor of Animal Physiology at the Radioisotope Department, Nuclear Research Center, Atomic Energy Authority, Egypt. She has authored many peer-reviewed articles on chronic diseases and is a referee, reviewer, and editorial board member for several international scientific journals. She obtained a Certificate of Excellence in international scientific research arbitration from Publons and was selected for inclusion in the Who's Who directory in 2014, 2015, and 2016.

# Contents

# Preface

Recycling is easy, valuable, and eco-friendly and reduces the need to extract raw materials from the environment. The environmental problems of waste accumulation go beyond irritating odors and unpleasant scenery to leakage of toxins and the creation of fertile environments for the spread of many diseases and epidemics. In addition, typical waste disposal operations increase carbon dioxide emissions through burning fossil fuels or through emissions of methane, ammonia, and sulfur dioxide waste directly, which contributes to the intensifying of global warming. The recycling process is complex because the process itself consumes energy and resources, but at the same time, it reduces the use of raw materials that save up to two million tons of carbon dioxide equivalent. Recycling ultimately plays a role in reducing emissions and protecting the climate by preserving natural environments.

Waste is currently the third greatest renewable energy resource worldwide, after solar and wind energy. Along with biomass energy, it contributes to more than half of the renewable energy used globally. This is why many countries have researched and developed plans to separate and recycle the garbage or convert it into compost. Now, due to the tremendous development in the science of solid waste management, more than half of the garbage is incinerated and converted into liquid or gaseous fuels. The extraction of energy from solid waste is an encouraging option for large cities due to the lack of space allocated for landfills and the high material and environmental costs of transporting garbage. In several industrialized countries, burning waste is one of the heat sources needed for heating and generating electric power. The emission of dust, acids, metals, and organic matter from old and modern incinerators is well monitored in most of the world's large cities. Plastic waste causes a lot of devastating environmental damage, especially when burned. Declining plastic recycling rates, along with an increase in plastic pollution on the Earth's surface and the oceans, is very concerning. Given the volume of waste on a global level due to the sharp increase in population numbers and changing lifestyles, the high cost of solid waste treatment technologies and the limited energy yield resulting from them are barriers that limit the spread of these technologies in developing countries. The steady increase in waste production rates represents an urgent global environmental problem, with dangerous environmental consequences, especially for major cities.

This book focuses on recycling strategies and technologies to find solutions to manage waste and benefit from it for the sustainability of the environment. Written by researchers and scientists in the field, it includes nine chapters that provide background on the culture of recycling and present alternatives to typical waste management methods. Each chapter is a comprehensive review of the positive impact of recycling and waste management on biological systems. The idea of converting waste into sources of energy is very interesting and represents a future challenge that must be focused on and developed by all possible means.

The chapters are organized into two parts covering important research aspects of recycling and waste management techniques. The first section, “Recycling, Waste Management and Sustainable Utilization”, includes six chapters.

- Chapter 1: “Biodegradable Waste”
- Chapter 2: “Alternatives for the Management of Industrial Forest Waste: Energy, Bioethanol, and Cellulose Pulp”
- Chapter 3: “Nuclear Waste Hazard Reduction”
- Chapter 4: “Revolutionary Plastic Mechanical Recycling Process: Regeneration of Mechanical Properties and Lamellar Structures”
- Chapter 5: “Recycling Gap, Africa’s Perspective for Sustainable Waste Management”
- Chapter 6: “Recycling Asphalt Pavements: The State of Practice”

The second section, “Circular Economy and Waste Challenges”, includes three chapters.

- Chapter 7: “Circular Economy: An Antidote to Municipal Solid Waste Challenges in Zambia”
- Chapter 8: “Design for Recycling of E-Textiles: Current Issues of Recycling of Products Combining Electronics and Textiles and Implications for a Circular Design Approach”
- Chapter 9: “Tesla’s Circular Economy Strategy to Recycle, Reduce, Reuse, Repurpose and Recover Batteries”

We wish to thank all the participating authors for their valuable contributions. We are also grateful to the staff at IntechOpen, especially Author Service Manager Ms. Dolores Kuzelj for her continuous assistance in finalizing this work.

**Hosam M. Saleh and Amal I. Hassan**
Egyptian Atomic Energy Authority,
Cairo, Egypt

Section 1

# Recycling, Waste Management and Sustainable Utilization

Chapter 1

# Biodegradable Waste

*Bukola Margaret Popoola*


## Abstract

Biodegradable wastes are waste materials easily degraded or broken down naturally by factors such as biotic (bacteria, fungi, plants, animals, etc.) and abiotic (pH, temperature, oxygen, humidity, etc.). This process enables complex substances to be broken down into simpler organic compounds which subsequently fade into the soil. This is a natural process that could be prolonged or rapid and poses little risks to the environment. These waste materials could be termed green waste; including food waste, paper waste, and biodegradable plastics such are found in municipal solid waste. Other examples of biodegradable wastes include sewage, manure, sewage sludge, human waste, waste from various slaughterhouses, hospital waste, dead animals, and plants. Biodegradable waste could be said to be recyclable or reused; furthermore, bio-waste recycling may also directly contribute to climate protection. They are generally known as useful waste. Recycling is one of the current waste management strategies having great benefits for the environment.


**Keywords:** recycling, biodegradable, biotic, bio-waste, environment

## 1. Introduction

Waste can be said to be an inevitable constituent arising as a consequence of domestic activities or industrial action. They have little or no value, due to the fact that they generally have no alternative use. Lack of adequate waste disposal system generates a great challenge for both the environment and human life. Waste can be divided into biodegradable and non-biodegradable waste. Non-biodegradable wastes are inorganic sources of waste that are not easily decomposed by natural agents, they can remain on the planet for hundreds of decades. They are sources of great damage to the ecosystem examples include plastics, batteries, glass, metal, medical waste, etc. However, many of them can be recycled to produce new products.

Biodegradable wastes are waste materials easily degraded or broken down naturally by factors such as biotic (bacteria, fungi, plants, animals, etc.) and abiotic (pH, temperature, oxygen, humidity, etc.). The process is such that complex organic matter is broken down into simpler organic compounds such as carbon dioxide, water, methane, or simple organic molecules by microorganisms and other living things, acting in composting, aerobic digestion, anaerobic digestion, or similar processes [1].

This is a natural process that could be prolonged or rapid and poses little risks to the environment. These waste materials could be termed green waste (any biological waste that can be broken down into compost); including food waste, paper waste, and biodegradable plastics such are found in municipal solid waste. Other biodegradable wastes include human waste, manure, sewage, sewage sludge, and slaughterhouse waste [1].

However, if these biodegradable wastes are not properly managed they could become sources of pollution, thereby impacting the health of the environment negatively. Current clean-up strategies including recycling biodegradable waste have endeavored to mitigate the detrimental impacts of such waste on the environment. This review chapter addresses the current methods (Recent Advances) in biodegradable waste management, which when adequately implemented, can reduce the impact of such waste on health and environment.

## 2. Types of biodegradable waste

Biodegradable waste can be commonly found in municipal solid waste as green waste, food waste, paper waste, and biodegradable plastics. Other biodegradable wastes include human waste, manure, sewage, and slaughterhouse waste.

1. Green Waste: This can also be referred to as "biological waste", it is known as any organic waste that can be composted. Its constituent is usually refusing from gardens such as leaves or grass clippings, and industrial or domestic kitchen wastes. Materials such as pine, hay, dried leaves, or straw are not considered green waste, they are termed "brown wastes" being rich in carbon, on the other hand, green wastes contain concentrations of nitrogen. Green waste can also be used the increment efficiency of several composting operations and can be introduced to soil to sustain local nutrient cycling. Many communities, especially in the United Kingdom, have initiated green waste recycling and collection programs in order to reduce the quantity of biodegradable materials in landfills [2].

2. Food Waste: There is rapidly increasing attention given to sustainable food and biodegradable waste management (FBWM) in the municipal solid waste management system, due to environmental challenges. The enormous amount of FBW generated in Japan, for instance, is a result of their preference for raw food; such as raw egg, raw vegetables, or raw fish or meat (sashimi, sushi, etc.). Fresh fruits in particular, as well as other food, population growth, and improvement in lifestyle and standards of living, have invariably introduced several logistical problems that result in massive amounts of FBW [3, 4].

   Over the years, food and biodegradable waste have become a serious challenge and concern for both the general public and the government [5–7]. However, this waste presents a great opportunity if it is put to better use.

   It is worth noting that food waste constitutes a fraction of the biodegradable waste which historically has gained less attention, in spite of the fact that it is the most likely waste stream that could contaminate other waste fractions. Moreover, it has been the major contributor to methane production in landfill [8].

3. Paper Waste: Paper waste causes severe problems in so many offices and industries over the world. Due to printing mistakes, billings, junk mails, and packaging, such paper could comprise approximately 70% of the total waste of a company. The recycling of paper is done by taking it to the recycling plant where it is firstly

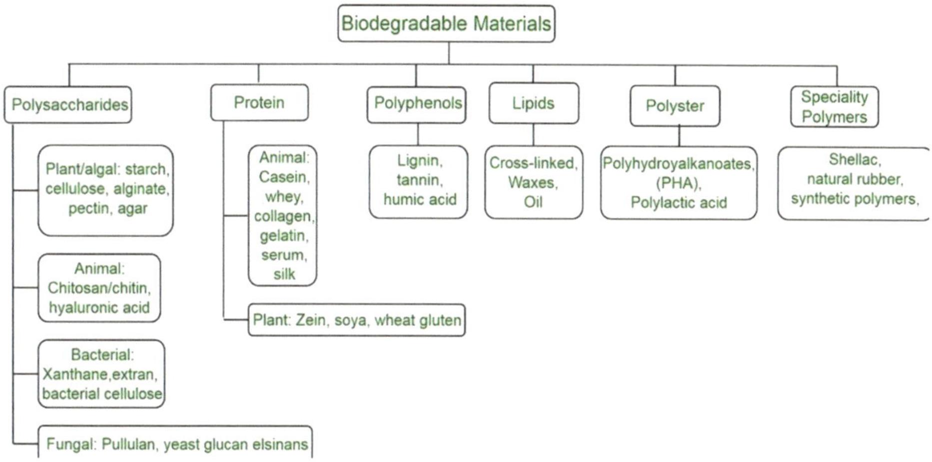


**Figure 1.**
*Biodegradable materials classification. Source: [9].*

separated and then the separated paper is cleaned and washed with soap in order for it to be broken down. After it is broken down, it is then exposed to heat after which it breaks down into cellulose. Recycling is an essential method to reduce pollution and cut back on waste accumulation. Notebooks, old newspapers, and used envelopes can be recycled. However, paper contaminated with food, stickers, and carbon paper cannot be recycled (**Figure 1**).

4. Biodegradable plastics: They are plastics commonly produced with petrochemicals, microorganisms, renewable raw materials, or combinations of all three. Biodegradable plastics can be decomposed by the activity of living organisms, usually microorganisms, into carbon dioxide, water, and biomass. Two types of biodegradable plastic are known: oxo-biodegradable (OBP) and hydro-biodegradable (HBP). They both begin degradation via a chemical process oxidation and hydrolysis respectively, which is then followed by a biological process. They both emit carbon dioxide as they degrade but the latter can also emit methane as well, it is also not recyclable unlike the former which is recyclable.

   These biodegradable plastics are used in various areas in our daily lives including: They make compostable waste collection bags, and trays, punnets for meat, vegetable, and fruits. The use of bioplastics to make pins, plates, screws, and materials for capsules and pills. They are also used for tea bags, air pillows, pencils, pens, bottles, mulch film, and sharpeners [10].

## 3. Harmful effects of biodegradable waste

Biodegradable wastes may pollute and impact the environment in the following ways:

- A large amount of microbial flora around the wastes is produced which may increase the risk of communicable diseases in humans, plants, and animals caused by microbes.

| Biodegradable Products | Period |
|---|---|
| Cotton rags<br>Paper | 1–5 months<br>2–5 months |
| Rope | 3–14 months |
| Orange peels | 6 months |
| Wool socks | 1–5 years |
| Cigarette butts | 1–12 years |
| Plastic-coated paper milk cartons | 5 years |
| Leather shoes | 25–40 years |
| Nylon fabric | 30–40 years |
| Plastic 6-pack holder rings | 450 years |
| *Source: [9].* | |

**Table 1.**
*Length of time for some commonly used products to biodegrade when they are scattered about as litter.*

- Bad odor on burning may be produced due to the emission of certain gases.
- Waste collection may lead to dungeons of garbage thus promoting the carriers and vectors like mosquitoes and rats to spread communicable diseases.
- Lack of biodegradable waste handling may lead to an adverse effect on climate. For instance, methane emission from anaerobic fermentation may result in the production of landfill gas (**Table 1**).

## 4. Management of biodegradable wastes

Waste management could be said to be procedures and measures necessary to manage waste from its commencement to its disposal stage [11]. This involves the collection, transport, treatment, and eventually waste disposal, at the same time monitoring and waste management regulation process and laws governing waste, economic mechanisms, and technologies. Waste management can also include the reduction of waste production.

Waste management is usually aimed at reducing the adverse effects of waste on human health, the environment, planetary resources, and esthetics. A great deal of waste management addresses municipal solid waste, which is largely generated by household, commercial and industrial activity. It is worth noting however that waste management practices are not consistent among countries (developing and developed nations); regions (rural and urban areas), and residential and industrial sectors can all take distinct approaches [12].

Ankidawa and Emmanuel [13], stated that biodegradable materials constitute approximately 70% of the urban waste stream in evolving economies. Disposal of refuse in an indiscriminate manner is a frequent occurrence in many villages and big cities across some developing countries, these waste decay, as well as the generated odors, pollute the environment where they are disposed of. This will subsequently lead to runoff of these decayed refuse into rivers and streams, by such means altering the quality of water sources from lakes, rivers, streams, etc. which could be deleterious to humans if ingested [13].

### 4.1 Waste handling and transport

Solid waste handling can be said to include the storage, collection, transportation, treatment, utilization, processing, or disposal of solid waste, or could mean any combination of such activities. Methods of waste collection vary widely among various regions and countries. For instance, the world was estimated to generate about 2.2 billion tonnes of solid wastes just in the year 2020 [14]. Of these generated wastes, a great percentage will be basically municipal solid waste (MSW). Certainly, the biosphere system will not be able to absorb and recycle such a large amount of wastes. Hence priority has to be given to the management of MSW for proper treatment and disposal in order to avoid a negative impact on the environment as well as on human health.

The domestic waste collection involves the handling in the local waste handling facilities. Domestic waste collection services are often provided by local government authorities, or by private companies for industrial and commercial waste. In some places in the underdeveloped countries, areas, especially those in less developed countries, do not have formal waste-collection systems.

Waste transportation means the movement of waste over specified areas by trains, tankers, barges, trucks, or other forms of vehicles. Train wrecks or traffic accidents can result in waste spills and releases of pollutants that may contaminate the air, soil, and water. There is also the possibility of waste being released while loading or unloading during transportation. Many citizens are concerned about the transportation of the waste through their communities and the risks involved. Some are as well bothered that the municipal waste from urban areas may be contaminated with substances that could contaminate local drinking water supplies or toxic chemicals.

Finally, the preparation of an Impact Assessment of a potential legislative proposal is of optimum importance. The objective is to look into various ways by which biodegradable waste is managed for instance in the EU, and to provide appropriate policy assessment options, such as the economic, social, and environmental impacts, as well as prospective opportunities/risks.

### 4.2 Segregation/separation of wastes

Waste or garbage most likely could be in the form of vegetable peels or fruits, wrapping materials, discarded objects, wasted food as household garbage, domestic sewage or chemicals, and fertilizers discarded and washed into rivers, etc. All these highlighted wastes can be segregated or separated into two basic components, as biodegradable and non-biodegradable components.

Contingent on the type of wastes, two garbage bins one for biodegradable wastes and the other for non-biodegradable wastes are usually advocated. This will help in easy sorting and recycling of wastes to make beneficial products. Green bins are designated for biodegradable wastes like fruit peels and vegetables, tea leaves, spoiled food, tissue paper, eggshells, hair, leaves, etc. Blue bins are designated for recyclable wastes like plastic waste, glass bottles, chocolate wrappers, old batteries, polythene bags, etc. However biodegradable waste could be managed by employing the following techniques:

I. Composting

Since biodegradable or organic wastes like vegetable peels, waste food, leaves, dead flowers, and eggshells can be recycled, they are converted into manure by burying them in compost pits. Composting can be said to be the act of burying recyclable organic wastes like leaves, waste food, vegetable peels, etc., in a compost pit. It is known to be a simple and almost effortless process of recycling. These wastes are biodegradable, this is a result of the action of small organisms like bacteria and fungi. It is also of importance to note that asides from bacteria and fungi acting in the compost, an earthworm called red worms (or red wigglers) can also act on wastes in the compost and degrade them.

Hence, composting, and the recycling of organic waste, including food waste and vegetation, decreases the amount of waste to be taken to the landfill, this is a sector that is growing rapidly. Ideally, compost residue has been reported to be a humus-like material, rich in organic matter, stable and sanitized, that is free from offensive odors arising from the composting process of segregating the biowaste collected [15]. Generally, recycling is thought to be beneficial to the environment, since leaving organic waste to decompose in landfills has a negative impact both economically and environmentally. Household waste that is commonly collected for recycling is the waste for composting, followed by paper and then glass [16] (**Table 2**). Although metal cans make up only 1% by weight of the material collected to be recycled, recycling them however offers high energy as well as material savings. Organic matter of biodegradable wastes is degraded microbiologically during composting, generating products containing stabilized carbon, nitrogen, and other nutrients in the organic fraction, however, the stability depends on the compost maturity [17].

II. Vermicomposting

The type of composting that is a result of decomposition via different species of worms, especially worms like the red wigglers, white worms, and other earthworms is called vermicomposting. They create a mixture of decomposing food waste or vegetables, vermicast (worm casting), and bedding materials. The worms usually break down the organic matter into nutrient-rich manure which increases soil fertility. Vermicompost can be said to be made in 3–4 weeks and it seems like loose soil-like material.

Vermicast, an end-product of the breakdown of organic matter by earthworms could also be referred to as worm castings, worm manure, worm humus, or worm feces. It has been demonstrated that these excreta contain lower levels of contaminants and a higher nutrient content than the organic materials before vermicomposting [18]. Furthermore, vermicomposting is a superb organic fertilizer that is rich in

| Material | % of household waste | Energy | Emissions | Raw material save/tonne recycled |
|---|---|---|---|---|
| Paper | 18 | 28–70% less | 95% less air pollutants | |
| Glass | 7 | 18% less | 30% less | 1.2 |
| Plastic | 7 | Up to 66% less | | 1.8 |
| Cans (Fe) | 3 | 70% less | 86% less | 2 |
| Can (Al) | 3 | 95% less | 95% less | 4 |

*Source: [16].*

**Table 2.**
*Impact of recycling for different materials.*

nutrients and soil conditioner, contains water-soluble nutrients and is an excellent, nutrient-rich organic fertilizer and soil conditioner [19]. It is also used in sustainable organic farming and gardening. It is worth noting that vermicomposting can be applied as well for sewage treatment. A variation of the process is vermidigestion or vermifiltration which is used to eliminate pathogens, organic matter, and oxygen demand from wastewater or directly from blackwater or flush toilets [20, 21].

III. Landfills

Sanitary landfills are generally used to dispose of non-hazardous solid wastes in an approach that reduces damage to human health and the environment. Before the evolution of sanitary landfills, solid wastes were frequently piled up on the ground in open-burning dumps, attracting rodents and insects as well as causing esthetic and public health issues.

Furthermore, landfills are usually large areas used for waste disposal. It is another method known to manage a vast amount of biodegradable waste. In a landfill, garbage is buried in such a way that it does not affect the environment negatively. Garbage buried inside landfills can be left there for a prolonged period as it decomposes very slowly. A landfill, when full, can be converted into a park. An example is Indraprastha Park.

Landfills have been traditionally used in handling large quantities of solid waste, being a low-cost method of waste management. As a result of rising in costs and reduction in land availability, many cities are focused on ways to reduce the quantity of solid waste dumped in landfills. For instance, in some cities, the government charges people for collection of garbage based on the size of the container collected. Hence, the bigger the can, the higher the cost.

This is planned to raise people's cognizance of the quantity of solid waste being generated as well as proffer an incentive for recycling. Programs to recycle plastic, paper, metal, and glass are currently being executed in various cities and countries with grand success. With the implementation of such programs, landfill areas can be anticipated to be readily available for a longer period of time. A case study is with the scenario in India, the management of waste is critical. The country holds 20% of the world's population and has only 2% of land space. There are inadequate spaces and locations to store waste. As a result, the primary focus should be on waste recycling and its reuse [22].

## 5. Uses and application of biodegradable waste

Anaerobic digestion used to treat biodegradable waste produces digestate a nutrient-rich solid material and biogas containing carbon dioxide and methane. However, for the biogas to burn to produce electricity, there is a need for further processing. This generated electricity can be used to power the plant. Alternatively, it could be used as a transport fuel. Biodegradable waste can also be used for composting or as a resource for electricity, fuel, and heat through the process of anaerobic digestion or incineration [23]. A classic example is that of Danish AIKAN and the Swiss Kompogas process of anaerobic digestion of biodegradable waste [24, 25]. While incineration can possibly recover the most energy, anaerobic digestion plants retain nutrients and make compost for soil amendment while still recovering some of the energy contained in the form of biogas. Kompogas for instance generated 27 million Kwh of electricity and biogas in 2009. The oldest of the company's lorries has achieved 1,000,000 kilometers driven with biogas from household waste for over 15 years [26].

## 6. Conclusion

Biodegradable wastes can be described as those wastes, whose source is typically of plant or animal origin that can be degraded by other living organisms such as microorganisms. They can be commonly found in municipal solid waste as green waste, food waste, paper waste, biodegradable plastics, etc. If not properly managed they could become sources of environmental pollution, resulting in sicknesses and diseases and possibly death to humans and other living organisms as well. There are various steps by which biodegradable waste could be managed such as the collection, transport, treatment, and eventually waste disposal.

Biodegradable waste is not accumulated but is used up in a short time, they become part of biogeochemical cycles and give back rapid turnover.

They can be used to produce renewable waste as a bio-energy resource (for instance biogas), manure, fertilizers, compost, and other substances after decomposition.

## Acknowledgements

I would like to acknowledge the staff and students of the department of Biological Sciences, Ajayi Crowther University, Oyo, Oyo State, Nigeria.

## Conflict of interest

"The authors declare no conflict of interest."

## Author details

Bukola Margaret Popoola
Ajayi Crowther University, Oyo, Nigeria

*Address all correspondence to: bm.popoola@acu.edu.ng

## References

[1] Rani R. Management of biodegradable waste. Journal of Biotechnology and Biomaterials. 2021;**11**(6):132. ISSN: 2155-952X

[2] Williams ID, Kelly J. Green waste collection and the public's recycling behaviour in the Borough of Wyre, England. Resources, Conservation and Recycling. 2003;**38**(2):139-159. DOI: 10.1016/s0921-3449(02)00106-4. ISSN 0921-3449

[3] Ministry of Agriculture, Forestry and Fisheries. Annual report on food, agriculture and rural areas in Japan FY 2007. Ministry of Agriculture, Forestry and Fisheries. 2008. Available from: http://www.maff.go.jp/e/annual_report/2007/pdf/e_all.pdf [Accessed: 2015-09-20]

[4] Takatsuki H. Waste problems and our lifestyle. Waste Management. 2013;**33**:2145-2146

[5] Gentil EC, Gallo D, Christensen TH. Environmental evaluation of municipal waste prevention. Waste Management. 2011;**31**:2371-2379

[6] Lebersorger S, Schneider F. Discussion on the methodology for determining food waste in household waste composition studies. Waste Management. 2011;**31**:1924-1933

[7] Wong L-F, Fujita T, Xu K. Evaluation of regional bioenergy recovery by local methane fermentation thermal recycling systems. Waste Management. 2008;**28**:2259-2270

[8] Gromez MGC, Grimes SM, Moore D. In-vessel composting of food waste – A catering waste management solution. CWRM. 2008;**9**(1):19-23

[9] GEEKSFORGEEKS. Biodegradable and non-biodegradable materials. 2022. Available from: https://www.geeksforgeeks.org/cdn.ampproject.org/biodegrdable and non-biodegradable materials [Accessed: 2022-08-19]

[10] Sciencing. What are the benefits of biodegradable plastic?. 2018. Available from: https://sciencing.com/benefits-biodegradable-plastic-22789.html [Accessed: 2022-08-20]

[11] United Nations Statistics Division. Environment statistics. Available from: unstats.un.org. Archived from the original on 17 March. 2017. Retrieved 3 March 2017

[12] Gary D. Waste management practices: Literature Review (PDF). Dalhousie University – Office of Sustainability. Archived (PDF) from the original on 1 February 2012. 2011. Retrieved 3 June 2022

[13] Ankidawa AB, Emmanuel N. Recycling biodegradable waste using composting technique. Journal of Environmental Science and Resources Management. 2012;**4**:40-49

[14] Christine Montgomery. World Bank, Solid Waste Management. 2022. Available from: https://www.worldbank.org. [Accessed: 2022-08-21]

[15] European Commission (EC). Working Document on Biological Treatment of Biowaste, 2nd Draft. 2001

[16] Postnote. Recycling household waste. Postnote recycling household waste. 2005;**252**:1-4

[17] Zwart K. Fate of C and N pools-experience from short and long term compost experiments. In: Anon, editor. Applying composts benefits and needs. Brussels Federal Ministry of Agriculture,

Forestry, Environment and Water Management, Austria and European Communities. 2003. pp. 77-86

[18] Ndegwa PM, Thompson SA, Das KC. Effects of stocking density and feeding rate on vermicomposting of biosolids (PDF). Bioresource Technology. 1998;**71**:5-12. DOI: 10.1016/S0960-8524(99)00055-3

[19] Coyne K, Knutzen E. The Urban Homestead: Your Guide to Self-Sufficient Living in the Heart of the City. Port Townsend: Process Self Reliance Series; 2008

[20] Xing M, Yang J, Wang Y, Liu J, Yu F. A comparative study of synchronous treatment of sewage and sludge by two vermifiltrations using an epigeic earthworm Eisenia fetida. Journal of Hazardous Materials. 2011;**185**(2-3):881-888. DOI: 10.1016/j.jhazmat.2010.09.103. ISSN 1873-3336. PMID 21041027

[21] Pilot studies for vermifiltration of 1000m 3day of sewage wastewater. Available from: www.academia.edu. Retrieved 2016-02-21. [Accessed: 2022-08-17]

[22] Khatabook. What is non-biodegradable waste?. 2021. Available from: https://khatabook.com/blog/what-is-non-biodegradable-waste/ [Accessed: 2022-08-17]

[23] National Non-Food Crops Centre. NNFCC report on Evaluation of Opportunities for Converting Indigenous UK Wastes to Fuels and Energy Archived at the Wayback Machine 20 July 2011. Available from: nnfcc.co.uk [Accessed: 2022-08-20]

[24] AIKAN website. Available from: aikantechnology.com [Accessed: 2022-08-20]

[25] Recycling chain Archived 2012-03-23 at the Wayback Machine. Available from: kompogas-utzenstorf.ch [Accessed: 2022-08-20]

[26] Gesundheit, Kraft und Energie für 2002. zuonline.ch. 3 January 2002. Archived from the original on 2 September 2002 [Accessed: 2022-08-20]

Chapter 2

# Alternatives for the Management of Industrial Forest Waste: Energy, Bioethanol, and Cellulose Pulp

*Leonardo Clavijo, Mairan Guigou, Norberto Cassella, Gastón Cortizo, Florencia Risso, Lucía Velazco, Mario Daniel Ferrari, Claudia Lareo and María Noel Cabrera*

## Abstract

Modern kraft pulp mills generate solid waste of 1–2% of incoming debarked wood. Given the size of these plants, with an annual production capacity of at least 1000,000 tons, each plant generates 20,000–30,000 dry tons of waste per year. The largest current use of these residues is for combustion in biomass boilers for steam and power generation. However, the conversion of biomass into biofuels and chemicals is gaining interest due to increasing demands for energy, limited sources of fossil fuels, and growing concerns about the environmental impact of greenhouse gas emissions. This chapter shows the laboratory-scale results of the use of eucalyptus wood wastes to obtain cellulose pulp by alkali pulping reinforced with hydrogen peroxide to obtain alkaline peroxide mechanical pulp or cellulosic bioethanol. Based on the results, an industrial-scale techno-economic analysis of the processes is presented and compared with current alternatives for energy generation.

**Keywords:** eucalyptus residues, APMP, bioethanol, valorization, industrial forest waste

## 1. Introduction

Although waste prevention is normally the first option to avoid later problems in waste management, forest industrialization generates a large amount of wood residue. In the mechanical transformation of wood (e.g., sawn wood, plywood), 30–50% of the logs are not part of the final product (e.g., bark, sawdust, under-sized logs, knots, chips, etc.). Depending on the size and characteristics of this waste and the integration of the wood industrial sector, it will normally be used to produce pulp, boards, or other value-added products, and it can be used as raw material for steam and power generation [1–4].

In modern pulp mills that apply the best available technologies, 1–2% of the input wood is discarded as pin chips and fines, which are formed during the chipping process and cannot be used to produce commercial cellulose pulp due to their small size [5]. For older mills, the discarded fraction could reach 5–10% [6]. These residues, as well as waste from mechanical transformation industries, are normally burned for steam and power generation.

Direct combustion of biomass in steam boilers is a mature, proven technology for converting biomass to energy. The technology choice largely depends on the specific size of the boiler. Newer solid-fuel-fired boilers with a capacity over 15 MW are fluidized bed boilers, moving grate boilers, or pulverized fired boilers [7, 8]. Small-scale cogeneration plants (with a capacity lower than 15–20 MW) are globally widespread, and their size is normally determined by the local availability of biomass. However, the high capital cost of electricity generation and the small price gap between biomass logistics costs and electricity prices often make them less competitive [7, 9].

The main objective of this chapter is to analyze an alternative to industrial forest residues, in the understanding that their disposal through the generation of electrical energy is a process that will no longer be profitable in the future since it has higher costs than electricity production from solar or wind energy.

Alternatively, lignocellulosic biomass constitutes a material with many possibilities for the development of bioproducts and/or biofuels. It mainly comprises cellulose, hemicelluloses, lignin, and a minor fraction of extractives and inorganic compounds. Pretreatments are required to separate components to use lignocellulosic materials. Within the different chemical pretreatments, alkali is used because it has several desirable characteristics. Alkaline methods mainly use noncorrosive chemical products, and in many cases, reagent recovery is feasible due to the consequent economic and environmental advantages.

One strategy for sustainable development is to establish an economy based on the use of biomass (a "bioeconomy")—a renewable resource that should be obtained through a production system that preserves the environment. This requires an increase in products' availability and their corresponding production technologies to simultaneously substitute fuels and materials derived from fossil sources ("economy decarbonization"). This concept can be developed through a biorefinery approach that comprises sustainable biomass processing into a portfolio of marketable products through the combination and integration of different processes. A biorefinery using lignocellulosic materials of forest origin to obtain biofuels and other products contributes to better meeting sustainable development goals related to energy, industry, innovation, infrastructure, and climate. Bioethanol production from lignocellulosic materials as a substitute for gasoline allows the increased incorporation of renewable energy in transport without too many changes in the current infrastructure and engines, facilitating the transition toward decarbonization of this sector.

Obtaining ethanol requires a biomass pretreatment that improves the accessibility and susceptibility of cellulose for enzymatic hydrolysis (cellulases) to glucose, preventing degradation of carbohydrates and the formation of byproducts' inhibitors of hydrolysis and fermentation processes. This allows for both the fractionation of hemicellulose components and the recovery of lignin. Alkaline pretreatment uses sodium, potassium, calcium, or ammonium hydroxides as reagents and is typically performed at lower temperatures than acid or hydrothermal pretreatment. Sodium and calcium hydroxide (lime) are the most common reactants (conc. 0.05–0.15 $g_{hidroxide}/g_{biomass}$), although recirculating ammonia percolation or fiber explosion/expansion treatments by ammonium are also common. Pretreatment with sodium or calcium hydroxide is normally carried out at temperatures between 20 and 130°C, with reaction times ranging from minutes to hours or days, depending on the working temperature [10–12]. In materials with a high lignin content, such as wood, this pretreatment is less efficient, so its use is more common in agricultural and agro-industrial waste [12].

In alkaline hydrolysis, fiber solvation occurs, producing swelling. The intermolecular ester bonds that cause the cross-linking of the chains are saponified, opening the structure of the hemicelluloses and lignin, and splitting the glycosidic bonds of the hemicellulose chain. Acetyl and uronic groups also react with alkali, separating them from the fiber. Depending on the applied conditions, the depolymerization of lignin molecules can occur by cleavage of the internal molecular bonds of the α and β-aryl ether type, which contributes to the degradation of lignin. All of these changes improve access to the cellulose chains, which become less crystalline with the treatment by reacting to the hydrogen bonds that link them. Under conditions of greater intensity, "peeling" reactions of the hemicellulose and cellulose chains may occur, causing a decrease in the degree of polymerization [10, 13, 14].

Among alkaline pretreatments, an alternative is the addition of an oxidizing agent (oxygen or hydrogen peroxide [$H_2O_2$]) to the alkali. The reaction mechanism is similar to alkaline hydrolysis, but it is aided by an oxidative effect on lignin, increasing its splitting from the lignocellulosic biomass. Alkaline peroxide treatments have faster kinetics at low temperatures, thus presenting technological and economic advantages over conventional alkaline treatments [10, 12].

Alkaline peroxide mechanical pulp (APMP) production is a particular case within alkaline treatments at low temperatures. It consists of treating wood with sodium hydroxide (NaOH), $H_2O_2$, and stabilizers at temperatures below 100°C, during which simultaneous pulping and bleaching occur. The pretreated material then goes through the mechanical defibration and refining stages [15, 16]. This type of pulping is widely used in China as a process for obtaining semi-chemical pulps from wood and nonwood materials [17].

In this work, the main destination of pulp is as a filler in the production of paper, particularly tissue. There is evidence that the substitution of kraft pulp for high-yield pulps of up to 30% improves the final product's properties, providing increases in bulk, tensile strength, bending stiffness, internal bond strength, and elasticity modulus [18].

Likewise, several reports have applied alkaline pretreatment to various types of biomasses [19, 20] to improve subsequent enzymatic hydrolysis for the production of bioethanol [21, 22] or to extract hemicelluloses prior to kraft pulping [23, 24]. More recently, the use of alkaline treatment for the extraction of hemicelluloses to prepare xylan films and the extraction of hemicelluloses from commercial pulps to produce nanocellulose have been reported [25, 26].

Once the lignocellulosic matrix is broken, it is necessary to hydrolyze the cellulose to glucose, which is fermented to produce bioethanol. Cellulose hydrolysis is preferably carried out enzymatically to minimize further inhibitions caused by the coproducts of chemical hydrolysis. The hydrolysis of cellulose into glucose through an enzymatic process involves several cellulases that act synergistically: 2 exo-β-glucanases or cellobiohydrolases (CBH: CBH I and CBH II), endo-β-glucanases (EG: EG I, EG II, EG III, and EG V) and a β-glucosidase (βG)) [27]. EG and CBH break down cellulose into smaller polysaccharides. EG first randomly breaks internal glycosidic bonds, then CBH I cleaves sugars from reducing ends, while CBH II cleaves cellulose toward nonreducing ends. This second stage occurs in the liquid phase and is mainly due to the hydrolysis of cellobiose to glucose by ß-glucosidase [24]. The enzyme dose needed to reach an acceptable percentage of glucose and ethanol yields strongly depends on the raw material and pretreatment used [11]. The kinetic effects on the enzymatic saccharification of lignocellulosic are largely related to the presence of polymers and nonionic surfactants, which are partially associated with the removal of lignin and cellulase stability and activation [28].

The microorganism used in fermentation must have certain conditions: high ethanol yield, high productivity, and tolerance to high concentrations of ethanol and sugars. It is also desirable for them to work in the presence of sugar degradation compounds, which are commonly fermentation inhibitors. Furthermore, it is expected that these microorganisms will have resistance to high temperatures and fermentable sugar. The most used microorganism for bioethanol production on an industrial scale is the yeast Saccharomyces cerevisiae, which efficiently ferments hexoses and disaccharides, such as sucrose and maltose. However, it cannot metabolize pentoses, such as xylose and arabinose, which are components of hemicelluloses [11].

The aim of this work was to compare three alternative uses for small-sized industrial forest residues. Therefore, an alkaline pretreatment reinforced with H2O2 was used for the subsequent production of bioethanol or bleached semichemical APMP. Based on the laboratory-scale results, an industrial-scale techno-economic analysis of the processes is presented (based on a plant located in Uruguay) and compared with current power generation alternatives.

## 2. Materials and methods

### 2.1 Raw materials

Pin chips from a local Eucalyptus pulp mill were received, dried and stored in airtight plastic boxes. The particle size distribution and chemical composition are presented in **Table 1**. Material below 0.50 mm was discarded.

### 2.2 Alkaline pretreatment

The process began with the mechanical impregnation of the pin chips with the pretreatment reagents in a device specially designed for this purpose. They were then transferred to a reactor where pretreatment occurred. The treated pin chips were

| Component | Amount (%) | Method | Particle size distribution | |
|---|---|---|---|---|
| Glucan | 42.0 ± 1.8 | NREL\TP-510-42,618 | Above 3.36 mm | 7.6% |
| Xylan | 14.1 ± 0.2 | | 3.36–1.40 mm | 43.7% |
| Acid insoluble lignin | 20.1 ± 1.2 | | 1.40–1.19 mm | 38.0% |
| Acid soluble lignin | 3.2 ± 0.1 | | 1.19–0.50 mm | 10.7% |
| Acetyl groups | 2.9 ± 0.2 | | Below 0.50 mm | 4.1% |
| Arabinan | 1.0 ± 0.1 | | | |
| Ash | 0.7 ± 0.1 | NREL\TP-510-42,622 | | |
| Extractives in ethanol and water | 4.3 ± 0.1 | TAPPI T 204 cm-97 | | |

**Table 1.**
*Raw material composition and particle size distribution.*

refined in a disk refiner, and the resulting pulp was divided into two groups. One part was used to produce the APMP. For this, a two-stage bleaching sequence was carried out using (1) ethylenediaminetetraacetic acid (EDTA) and (2) $H_2O_2$. Then, the papermaking properties of the obtained pulp were determined. The pulp obtained from refining was subjected to an enzymatic hydrolysis process, and the liquid obtained was fermented to produce bioethanol. This process is outlined in **Figure 1**.

According to Kruzolek [16], pressure is essential to achieving proper impregnation. For this, a device was designed and built to allow the pin chips to be immersed in water and reagents. By applying pressure, the occluded air in the pores was eliminated. The device was then returned to atmospheric pressure, achieving a "sponge effect" by expanding the compressed material and simultaneously absorbing the impregnated liquor. The apparatus is shown in **Figure 2**. The reagents (NaOH and $H_2O_2$) and water were externally heated to 80°C in a container. The amount of water was calculated to achieve a liquid-to-dry solid ratio of 7. The device, with the pin chips inside, was placed in a container within the press. Pressure was applied until it reached 0.5 MPa, held for 10 minutes, and then released.

Once the impregnation process was complete, solids and liquids were transferred to a 3 L Parr reactor (model 4522 M) and instrumented with temperature control and agitation, where the process temperature was reached using a defined heating ramp. The pressure in the reactor was maintained at 0.8 MPa using nitrogen gas.

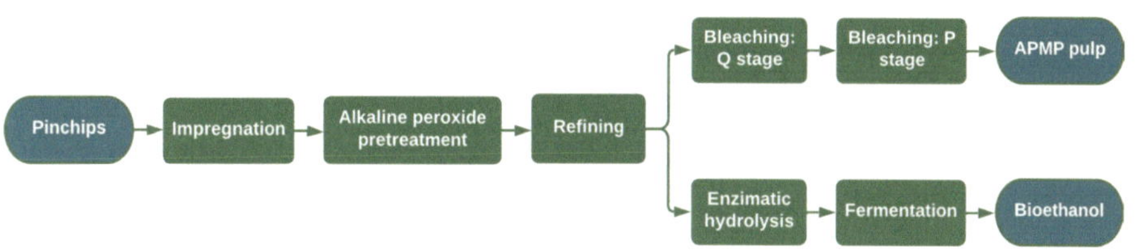


**Figure 1.**
*Schematic description of the process.*

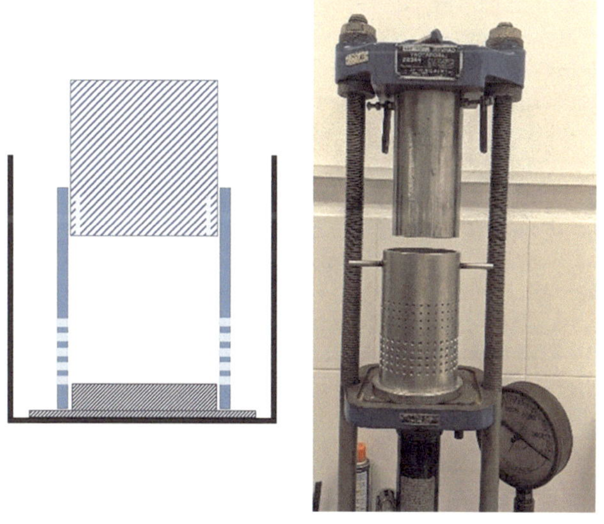

**Figure 2.**
*Impregnator and press are used to remove the air inside the pin chips.*

For pretreatment in the Parr reactor, preliminary trials were carried out following a factorial experimental design (Series A) using the following parameters: extraction temperature (50–100°C), time at that temperature (30–90 min), and doses of soda (8–12%) and peroxide (4–6%) used. In the design, a 2:1 (m:m) ratio of $NaOH:H_2O_2$ was applied. A constant level of 0.2% diethylenetriaminepentaacetic acid (DTPA) was used as a chelator. The function of DTPA is to remove heavy metals from the solution, particularly $Mn^{2+}$, $Cu^{2+}$, and $Fe^{3+}$, to prevent the decomposition of $H_2O_2$. According to Kruzolek [16], the colored extractives present in eucalyptus also react with this chelating agent.

After the extraction, the material was drained and centrifuged at 1000 rpm, which separated the solid fraction from the liquid. The solid fraction was then extensively washed with deionized water until a colorless filtrate was obtained.

In the solid fraction, the extraction yield was determined by considering the initial dry mass and the dry mass obtained after extraction, which was later refined in a disc refiner to produce semi-chemical pulp or to increase the accessibility to the cellulose chains to favor enzymatic hydrolysis. Refining was performed in a disc refiner (Kumagai Riki Kogyo Co., Ltd., Tokyo, Japan) with 30-cm diameter discs and a gap of 0.1 mm. The net yield after refining was also determined.

Brightness was measured (ISO 2470-1:2016) in the refined pulp. The pulps with the highest net yield and brightness were characterized in terms of carbohydrate and lignin content according to the NREL TP-510-42,618 procedure. The liquid fraction (liquor) obtained after alkaline peroxide treatment was characterized in terms of pH and carbohydrates, lignin, and biomass degradation product content (acetic and formic acids, furfural, and hydroxymethylfurfural (HMF)) (NREL TP-510-42,623). Residual peroxide was measured by iodometry.

This experimental design was evaluated in terms of extraction yield, brightness after refining, and enzymatic hydrolysis efficiency.

With approximately 40% of the design completed, the yields obtained were extremely low, and the enzymatic hydrolysis efficiency was far from the expected value. Therefore, and in accordance with the bibliographic review focused on bioethanol production [29, 30], more exhaustive extraction conditions were used (Series B). For these assays, the $NaOH:H_2O_2$ ratio was adjusted to 1:1 (m:m). Initial tests were carried out at 75°C for 90 min in the pretreatment stage, varying the NaOH charge between 5% and 45%. For different soda charges (between 8% and 30%), the processing time was increased while maintaining the temperature at 75°C.

## 2.3 Pulp bleaching and papermaking properties

To produce cellulose pulp, the solids were bleached using a two-stage sequence: the chelating stage (Q) and the $H_2O_2$ stage (P). The purpose of the first stage was to remove metals to avoid the catalyzed decomposition of $H_2O_2$ [15]. This was performed at acidic pH, using EDTA as a chelating agent. The aim of the second stage was to oxidize chromophoric compounds derived from lignin fragments into colorless structures. The bleaching power of $H_2O_2$ is highest at a pH between 10 and 11. The Q stage was performed at 65°C for 60 min using charges of 0.4% EDTA and 0.8% sulfuric acid (initial pH = 3). The P stage was performed at 80°C for 150 min using charges of 0.3% NaOH, 0.375% $H_2O_2$, 0.02% magnesium sulfate, and 0.02% DTPA.

After bleaching, some pulps were selected based on the global yield and the final brightness obtained. The papermaking properties of these pulps were determined.

To determine the papermaking properties, pulps were refined using a PFI refiner at 4000 revolutions (ISO 5264-2). In refined and unrefined pulps, drainability in terms of Canadian standard freeness (TAPPI T227 om-17) was determined. Pulp sheets were then prepared and conditioned according to ISO standards (ISO 5264-2, ISO 5269-2, ISO 187). On those sheets, brightness, opacity (ISO 2471:1998), basis weight, tear and tensile index, and tensile energy absorption (TEA) were measured according to TAPPI T410 om-08, TAPPI T414-om86, and TAPPI T494-om04 standards, respectively.

### 2.4 Enzymatic hydrolysis

The refined pulps were enzymatic hydrolyzed to obtain a liquid fraction rich in glucose, to be fermented for bioethanol production. All assays were performed under the same enzymatic hydrolysis conditions. A commercial enzymatic complex (Cellic CTec 2 (Novozymes)) was used. Enzyme dosage was 15 $FPU/g_{glucan}$ with a solid loading of 15% (w/w). The enzymatic hydrolysis was performed in Erlenmeyer flasks, pulps were suspended in a 0.05 M citric acid buffer solution (pH = 4.8) at 48°C for 96 h with orbital agitation at 150 rpm. Samples were taken at times 0 and 96 h. Tests were carried out by duplicate.

## 3. Results and discussion

### 3.1 Alkaline pretreatment

**Figure 3** shows the aspects of the raw material, alkaline-treated solid, alkaline-extracted liquor, and refined pulp.

As expected, the solid obtained after pretreatment was brighter than the original raw material. This demonstrates the effect of $H_2O_2$. However, the liquor obtained presented a dark color, similar in visual appearance to that obtained in treatments without peroxide.

**Table 2** shows the liquid and solid fraction results after pretreatment and refining assays (Series A).

The extraction yields for the alkaline peroxide pretreatment varied between 78.5% and 90.5%. However, no clear trend was observed within the parameters studied. These yields are comparable to those obtained by Liu et al. [31], who worked with aspen chips under similar conditions (temperature: 50–100°C, time: 10–60 min, soda

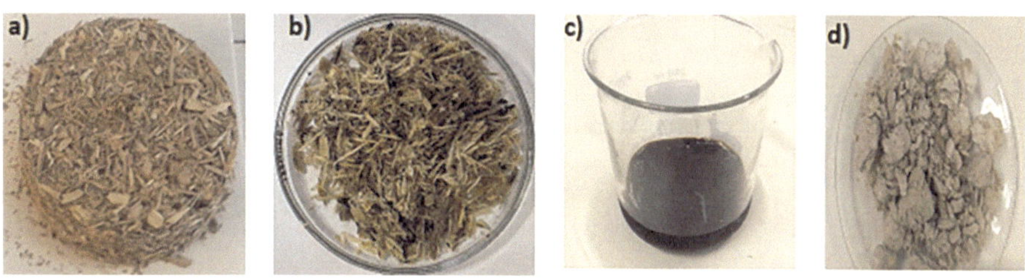


**Figure 3.**
*a) Eucalyptus pin chips used as raw material; b) solid fraction after alkaline-peroxide treatment; c) liquid fraction obtained after alkaline-peroxide treatment; d) pulp obtained after mechanical refining process.*

| Condition applied | | | | Solid fraction | | | Liquid fraction | | | |
|---|---|---|---|---|---|---|---|---|---|---|
| T (°C) | Time (min) | NaOH (%) | $H_2O_2$ (%) | Extraction yield (%) | Yield after refining (%) | Brightness (% ISO) | pH | Lignin (%) | XS (%) | Acetyl groups (%) |
| 50 | 30 | 10 | 5 | 90.5 ± 0.1 | 82.0 ± 0.1 | 49.7 ± 0.5 | 11.9 | 2.4 ± 0.1 | 0.7 ± 0.1 | 3.6 ± 0.1 |
| 50 | 30 | 8 | 4 | 85.1 ± 0.1 | 80.5 ± 0.1 | 48.3 ± 0.5 | 10.9 | 2.0 ± 0.1 | n.d. | n.d. |
| 50 | 60 | 8 | 4 | 89.5 ± 0.1 | 85.7 ± 0.1 | 50.2 ± 0.6 | 10.8 | 2.2 ± 0.1 | 0.5 ± 0.1 | 3.5 ± 0.1 |
| 50 | 90 | 12 | 5 | 86.6 ± 0.1 | 78.8 ± 0.1 | 48.6 ± 1.0 | 12.2 | 2.7 ± 0.1 | 0.9 ± 0.1 | 3.9 ± 0.1 |
| 50 | 90 | 8 | 4 | 87.8 ± 0.1 | 76.8 ± 0.1 | 53.4 ± 0.9 | 10.6 | 2.0 ± 0.1 | 0.5 ± 0.1 | 3.1 ± 0.1 |
| 75 | 30 | 10 | 5 | 90.2 ± 0.1 | 81.5 ± 0.1 | n.d. | 9.4 | 1.5 ± 0.1 | n.d. | n.d. |
| 75 | 30 | 8 | 4 | 78.4 ± 0.1 | 82.5 ± 0.1 | 48.8 ± 0.5 | 10.0 | 2.4 ± 0.1 | 0.5 ± 0.1 | 3.3 ± 0.1 |
| 75 | 60 | 8 | 4 | 85.1 ± 0.1 | 82.9 ± 0.1 | 50.2 ± 0.6 | 10.0 | 2.6 ± 0.1 | 0.7 ± 0.1 | 3.5 ± 0.1 |
| 75 | 60 | 10 | 5 | 78.5 ± 0.1 | 77.5 ± 0.1 | 48.9 ± 0.3 | 11.1 | 2.5 ± 0.2 | 0.2 ± 0.1 | 4.3 ± 0.1 |
| 75 | 90 | 8 | 4 | 89.1 ± 0.1 | 79. 6 ± 0.1 | 49.5 ± 0.6 | 9.6 | 1.9 ± 0.1 | 0.3 ± 0.1 | 2.3 ± 0.1 |
| 75 | 90 | 10 | 5 | 87.5 ± 0.1 | 81.3 ± 0.1 | 47.6 ± 0.3 | 11.1 | 2.9 ± 0.2 | 0.8 ± 0.1 | 3.8 ± 0.1 |
| 100 | 60 | 12 | 6 | 86.5 ± 0.1 | 71.2 ± 0.1 | 50.9 ± 0.3 | 9.1 | 2.2 ± 0.1 | n.d. | n.d. |
| 100 | 90 | 10 | 5 | 86.8 ± 0.1 | 83.0 ± 0.1 | 46.7 ± 1.2 | 10.4 | 2.7 ± 0.2 | 1.0 ± 0.1 | 3.9 ± 0.1 |

**Table 2.**
*Solid and liquid fraction characteristics after alkaline pretreatment and mechanical refining of the Serie A assays. (XS: Xylosaccharides; n.d.: Not determined).*

load 1.5%–6.7%, and NaOH:$H_2O_2$ ratio 2:1), obtaining extraction yields between 85% and 94%.

All liquors presented a pH higher than neutrality. The final pH of the liquors arose from the soda charge applied in each condition and the acetyl and/or uronic groups released during extraction, which partially neutralized the hydroxyl groups. Consequently, the highest pH in the liquor was achieved under low-severity pretreatment and a high soda charge.

Pulp brightness was between 46.7% and 53.4% ISO. The low variability detected among these values did not support the determination of a general trend. Kruzolek [16] reported brightness values in the range of 55.4–64.9% ISO when working with a smaller charge of soda but a higher NaOH:$H_2O_2$ ratio using *Eucalyptus grandis* chips. Moreover, Latibari et al. [32], when working with Paulownia wood at 70°C for 100 min and charges of NaOH and $H_2O_2$ between 1.5% and 5.0% each, reported an ISO brightness between 67.5%–69.0%. In this case, the highest value reached for brightness is explained by the fact that Paulownia wood has 12.1% of extractives in ethanol: acetone because these extractives are normally colored and easily removed in an alkaline medium during pretreatment before bleaching.

Regarding the liquid fraction, xylosaccharides (from the hemicellulose fraction in the raw material) and lignin content were too low to be used for producing value-added products.

The results obtained from the Series B assays are reported in **Table 3**.

When working at 75°C for 90 min, the extraction yield decreased linearly with the soda charge ($r^2$ = 0.97). Variations in the other parameters did not show a clear trend. Despite the large increase in the dose of soda and peroxide, the pulp's brightness hardly improved compared to the results obtained in Series A. Xylosaccharide content in the liquid fraction increased as much as 5.9%, which implies a removal of almost 40% of the Xylan present in the raw material. Li et al. [33] carried out a treatment using corn cobs with 50% $H_2O_2$ (pH 11.5) at 30°C for 24 h and extracted 56% of the initial hemicelluloses and 92% of the initial lignin. Gupta and Lee [34], when working at 60°C for 24 h with a soda and $H_2O_2$ charge of 50%, reported extraction of 40% of the xylosaccharides originally present in poplar wood. Hemicellulose extraction yields differ among authors and depend on the materials and conditions used. The latter tend to have longer reaction times, a high soda and peroxide charge, and low temperatures. The extracted xylosaccharides can be recovered by precipitation with ethanol and used for biofilm production [35, 36].

### 3.2 Pulp bleaching and papermaking properties

The bleaching procedure was applied to all pulps, and the nine pulps with the highest values of brightness and net yield were selected. These pulps were refined, and the papermaking properties of the refined and unrefined pulps were determined.

The mechanical and optical properties of the selected bleached pulps are shown in **Table 4**.

As can be expected, sheet density increased when the pulp was refined, making fibers more flexible and increasing the bonding capacity and the bonding area. Additionally, external fibrillation collaborates to increase interfiber bonding formation [15, 37]. The sheet density in this work was higher than the density reported in other studies working with eucalyptus chips [15, 16] or chips of other hardwood species [32]. The use of small wood increased the external fibrillation during refining, which

| Conditions applied | | | | Solid fraction | | | Liquid fraction | | | |
|---|---|---|---|---|---|---|---|---|---|---|
| T (°C) | Time (min) | NaOH (%) | $H_2O_2$ (%) | Extraction yield (%) | Yield after refining (%) | Brightness (% ISO) | pH | Lignin (%) | XS (%) | Acetyl groups (%) |
| 75 | 90 | 5 | 5 | 88.4 ± 0.1 | 78.4 ± 0.1 | n.d. | 8.9 | 1.5 ± 0.1 | 0.6 ± 0.1 | 3.2 ± 0.3 |
| 75 | 90 | 10 | 10 | 83.3 ± 0.1 | 78.7 ± 0.1 | n.d. | 11.3 | 2.4 ± 0.1 | 1.3 ± 0.1 | 3.1 ± 0.1 |
| 75 | 90 | 15 | 15 | 80.1 ± 0.1 | 51.2 ± 0.1 | n.d. | 12.8 | 2.5 ± 0.1 | 2.5 ± 0.1 | 3.4 ± 0.2 |
| 75 | 90 | 30 | 30 | 72.2 ± 0.1 | 66.8 ± 0.1 | 51.6 ± 0.5 | 12.7 | 3.2 ± 0.2 | 4.4 ± 0.2 | 3.5 ± 0.3 |
| 75 | 90 | 45 | 45 | 71.9 ± 0.1 | 61.2 ± 0.1 | 58.9 ± 0.9 | 12.5 | 3.9 ± 0.2 | 5.9 ± 0.2 | 3.5 ± 0.2 |
| 75 | 90 | 60 | 60 | 65.6 ± 0.1 | 53.7 ± 0.1 | 45.2 ± 0.4 | 12.3 | 4.3 ± 0.2 | n.d | 3.5 ± 0.3 |
| 100 | 240 | 8 | 8 | 85.9 ± 0.1 | 56.3 ± 0.1 | n.d. | 6.9 | 2.4 ± 0.1 | 2.7 ± 0.3 | 3.2 ± 0.2 |
| 100 | 240 | 12 | 12 | 78.1 ± 0.1 | 67.0 ± 0.1 | n.d. | 9.6 | 2.8 ± 0.1 | 3.6 ± 0.3 | 3.6 ± 0.3 |
| 75 | 120 | 8 | 8 | 88.3 ± 0.1 | 58.8 ± 0.1 | n.d. | 7.4 | 1.8 ± 0.1 | 2.1 ± 0.1 | 3.5 ± 0.2 |
| 75 | 120 | 12 | 12 | 93.6 ± 0.1 | 48.6 ± 0.1 | n.d. | 10 | 2.6 ± 0.1 | 1.9 ± 0.1 | 3.5 ± 0.1 |
| 75 | 240 | 16 | 16 | 82.6 ± 0.1 | 66.6 ± 0.1 | n.d. | 12.8 | 2.6 ± 0.1 | 0.5 ± 0.1 | 3.9 ± 0.3 |

**Table 3.**
*Solid and liquid fraction characteristics after alkaline pretreatment and mechanical refining of the Serie B assays. (XS: Xylosaccharides; n.d.: Not determined).*

DOI: http://dx.doi.org/10.5772/intechopen.107528

| T (°C) | Time (min) | NaOH (%) | $H_2O_2$ (%) | Beating rev. (rpm) | CSF (mL) | Sheet density (kg/m$^3$) | Brightness (% ISO) | Opacity (%) | Tensile Index (Nm/g) | Tear Index (Nm$^3$/g) | TEA (J/m$^2$) |
|---|---|---|---|---|---|---|---|---|---|---|---|
| 50 | 60 | 8 | 4 | 0 | 349 | 405 | 64.5 | 93.6 | 15.9 | 1.9 | 0.08 |
| | | | | 4000 | 107 | 546 | 63.7 | 93.2 | 30.2 | 1.8 | 1.1 |
| 50 | 90 | 8 | 4 | 0 | 200 | 452 | 73.3 | 89.6 | 23.9 | 2.18 | 0.2 |
| | | | | 4000 | 76 | 595 | 72.5 | 90 | 37.5 | 2.04 | 0.34 |
| 75 | 30 | 8 | 4 | 0 | 302 | 453 | 75.6 | 86.3 | 24 | 2.32 | 0.2 |
| | | | | 4000 | 72 | 609 | 75 | 86.9 | 42.7 | 2.3 | 0.46 |
| 75 | 90 | 30 | 30 | 0 | n.d | n.d. | n.d. | n.d. | n.d. | n.d. | n.d. |
| | | | | 4000 | 57 | 786 | 74 | 78.5 | 42.6 | 1.44 | 0.34 |
| 75 | 90 | 45 | 45 | 0 | n.d | n.d. | n.d. | n.d. | n.d. | n.d. | n.d. |
| | | | | 4000 | 37 | 777 | 73.8 | 77.5 | 41.6 | 1.2 | 0.31 |
| 75 | 120 | 8 | 8 | 0 | 340 | 455 | 73.5 | 84.1 | 25.5 | 2.86 | 0.21 |
| | | | | 4000 | 81 | 623 | 72.9 | 84.5 | 47.2 | 2.93 | 0.72 |
| 75 | 240 | 16 | 16 | 0 | 339 | 526 | 72.6 | 83 | 35.6 | 3.57 | 0.36 |
| | | | | 4000 | 79 | 681 | 71.1 | 82.4 | 60.0 | 3.36 | 0.88 |
| 100 | 60 | 12 | 6 | 0 | 529 | 410 | 76.2 | 82.7 | 15.5 | 2.41 | 0.08 |
| | | | | 4000 | 120 | 624 | 75.4 | 84.6 | 46.8 | 2.87 | 0.55 |
| 100 | 90 | 10 | 5 | 0 | 300 | 518 | 70.4 | 86.9 | 35.3 | 3.23 | 0.43 |
| | | | | 4000 | 91 | 654 | 69.2 | 86.2 | 52.4 | 3.26 | 0.74 |

**Table 4.**
*Paper properties of selected bleached pulps at different degrees of refining.*

increased the sheet density. Notably, the density increased with the chemical charge, producing more fibrillation and bulkier pulp.

Brightness improved after bleaching, but the brightness of the commercial pulps was generally higher. However, it is possible to obtain pulp with a higher final brightness when using a longer residence time in the P stage, carrying out this stage in two steps, or with higher chemical doses. Opacity was high for almost all bleached pulps, which is good for paper applications.

Tensile strength is useful in determining resistance to breakage. It is associated with the strength, length, and bonding of the fibers, which are also dependent on the degree of refinement. The tensile strength of the bleached pulps refined at 4000 rpm was high for all the conditions tested, reaching a tensile index of 60.0 Nm/g. The tear index for the refined pulps was also good, in accordance with the values reported by other authors [38–40].

Of the nine bleached pulps tested, two showed the best papermaking properties. The conditions for these pulps were as follows: (1) 75°C for 240 min with a dose of 16% soda and 16% peroxide and (2) 100°C for 90 min with a dose of 10% NaOH and 5% peroxide. The first pulp had slightly better properties. However, the increase in chemical loads and the high residence time did not justify the use of these conditions at the industrial level. In addition, the second treatment had a noticeably higher yield (83.3% vs. 66.6%), so a greater amount of pulp was obtained from the same amount of pin chips. Consequently, the conditions selected to scale up the process were 100°C for 90 min and a load of 10% NaOH and 5% peroxide.

### 3.3 Enzymatic hydrolysis

The chemical composition of the pretreated solids used and their behavior in enzymatic hydrolysis are listed in **Table 5**.

The solid pretreated with $NaOH/H_2O_2$ presented a maximum enzymatic hydrolysis efficiency of 23% when 10%, 12%, and 30% soda loads were used, and then its value decayed 1.5 times when 45% soda was used. By adding 6.25% $H_2O_2$, the efficiency increased from 30–40% and from 40–49% by the cellular structure alteration due to hemicellulose (49%) and lignin (26%) removal, which left the cellulose more accessible to the enzyme [41]. In this work, only slightly lower hydrolysis efficiencies (23%) were obtained by pretreating eucalyptus with NaOH (10)/$H_2O_2$ (5) at 75°C for 90 min, despite the fact that lignin removal was much lower (2.8%). Bagasse lignin is more easily solubilized with alkaline solutions due to the presence of lignin H (which contains p-hydroxyphenyl in its structure), which is not present in the wood [42]. Alvarez-Vasco and Zhang (2017) used different temperatures (25–180°C), reaction times (30–90 min), and $H_2O_2$ concentrations (4–10%) with spruce and observed that temperatures below 140°C were not effective in hydrolyzing cellulose (efficiency less than 25%) [43].

An increase in peroxide charge or reaction time worsened the enzymatic hydrolysis of the cellulose process. Several authors have reported low yields of enzymatic hydrolysis in lignocellulosic materials pretreated with the addition of alkalis, which can be attributed to the formation of pseudo-lignin or the solubilization of the amorphous fractions of cellulose, leaving more crystalline zones in the solid. Pseudolignin acts as a barrier, reducing the accessibility of the enzymes or retaining toxic compounds (furfural, HMF, phenolic compounds), making it difficult to remove them during washing [44–51].

| Conditions applied | | | | Solid composition (%) | | | Hydrolysis parameters at 96 h | |
|---|---|---|---|---|---|---|---|---|
| T (°C) | Time (min) | NaOH (%) | $H_2O_2$ (%) | Glucan | Xylan | Lignin | Glucose (g/L) | Hydrolysis efficiency (%) |
| 100 | 60 | 12 | 6 | 55.0 ± 0.8 | 12.0 ± 0.1 | 29.4 ± 4.4 | 14.1 ± 1.0 | 14.2 ± 1.0 |
| 75 | 90 | 5 | 5 | 51.4 ± 1.4 | 16.5 ± 0.5 | 26.5 ± 1.7 | 6.7 ± 0.3 | 5.2 ± 0.2 |
| 75 | 90 | 8 | 4 | 54.0 ± 1.3 | 12.2 ± 0.3 | 30.0 ± 3.9 | 9.7 ± 0.5 | 9.9 ± 0.1 |
| 75 | 90 | 10 | 5 | 55.8 ± 1.4 | 12.1 ± 0.2 | 32.5 ± 5.5 | 22.3 ± 1.0 | 23.3 ± 3.3 |
| 75 | 90 | 12 | 12 | 48.7 ± 3.6 | 18.4 ± 2.1 | 28.1 ± 0.1 | 21.4 ± 1.5 | 22.9 ± 2.8 |
| 75 | 90 | 30 | 30 | 51.0 ± 0.1 | 10.7 ± 0.0 | 30.2 ± 3.0 | 22.0 ± 0.5 | 23.3 ± 1.8 |
| 75 | 90 | 45 | 45 | 53.1 ± 0.4 | 9.4 ± 0.3 | 26.4 ± 0.2 | 15.5 ± 0.9 | 15.8 ± 0.7 |
| 75 | 90 | 60 | 60 | 54.9 ± 0.1 | 9.9 ± 4.0 | 29.5 ± 1.3 | 9.9 ± 1.3 | 10.5 ± 1.2 |
| 75 | 240 | 30 | 30 | 54.6 ± 0.3 | 14.9 ± 0.2 | 24.5 ± 4.0 | 14.5 ± 3.4 | 15.1 ± 3.4 |
| 75 | 480 | 30 | 30 | 47.6 ± 1.3 | 12.3 ± 0.1 | 26.5 ± 0.3 | 12.2 ± 1.2 | 14.3 ± 2.0 |

**Table 5.**
*Solids composition before enzymatic hydrolysis, glucose content, and hydrolysis efficiency.*

The results obtained for the enzymatic hydrolysis of the alkaline peroxide-treated solids (based on the final concentration of glucose and enzymatic efficiency) were low. The low concentration of glucose obtained made it pointless to proceed with subsequent fermentation. Simultaneous hydrolysis and fermentation tests also had poor results (data not shown).

## 4. Techno-economic analysis

For the different processes, an economic analysis was carried out considering the annual profitability of each option. The revenues from product sales were equalized to the annual production costs (including the amortization of the investment), and the maximum price at which biomass could be purchased was calculated. This option was preferred to the classic analysis through the internal rate of return and the net present value because the investments required for the options listed below were quite different.

An economic comparison was performed for two options: (1) the installation of a power boiler plant to produce energy to be sold to the national grid and (2) the production of APMP to be sold to papermaking companies. To this end, the best conditions found in the experiment were used as working conditions. Due to the lower enzymatic hydrolysis efficiencies obtained, bioethanol production was not analyzed.

Both analyses were performed considering 50,000 dry tons/year of eucalyptus pin chips, with a water content of 45%, which corresponds with the number of pin chips generated for a modern kraft eucalyptus pulp mill with a capacity of 2.1 ADT[1]/year. The prices used correspond to a plant located in Uruguay.

[1] ADT stands for Air Dry Ton and is commonly used in the pulp industry as a unit of measure for the pulp produced. It represents pulp with a moisture content of 10%; therefore, 1 ADT is equivalent to 0.9 ODT (Oven Dry Ton).

### 4.1 Techno-economic analysis for power generation alternative

To calculate the power boiler size (MW), Eq. (1) was used, considering a boiler efficiency of 85% and an efficiency in the vapor cycle of 25% (to convert steam to electricity) [52].

$$Biomassconsumption = \frac{Boilerpower}{Boilerefficiency * LHV} \quad (1)$$

A lower heating value of 11.3 MJ/kg was considered [53]. The values applied for the economic analysis are listed in **Table 6**.

Considering utilities equal to zero to calculate the maximum price of biomass and considering the energy selling price in the range of 50–70 MWh, different scenarios can be analyzed. The results are shown in **Table 7**.

In the first two cases, the price of biomass was quite low and probably would not cover the cost of biomass handling. However, for a scenario where the energy price reached 70 USD/MWh, which could be possible in Uruguay, the biomass price increased to 16.3 USD/ton, which is an attractive value.

### 4.2 Techno-economic analysis for an APMP mill

Considering the yields obtained in the experiment for a biomass availability of 50,000 ADT/year, the size of the APMP mill turn should be 41,500 ADT/year. The values applied for the economic analysis are listed in **Table 8** [52, 54].

Amortizable capital costs include yard improvements, machinery, equipment, buildings, utility services, and electrical and piping systems.

| **Investment (USD[2])** | **13.850.000** |
|---|---|
| Operation | 24 hours/330 days/year |
| Life of project | 10 years |
| Installed kW cost[3] | 2700 USD/kWh |
| Connection to the national grid cost (USD) | 350.000 USD [53] |
| Yard improvements | 1% of fixed capital |
| Operation & Maintenance | 4% of fixed capital |

*[2]USD means Dollar USA.[3]Based on [53] and estimating the cost by scaling using the "six-tenths factor rule" [52].*

**Table 6.**
*Economic analysis parameters for power generation.*

| **Energy sale price (USD/MWh)** | **Biomass maximum purchase price (USD/dry ton)** |
|---|---|
| 50 | 0.43 |
| 60 | 8.4 |
| 70 | 16.3 |

**Table 7.**
*Result of economic analysis for power generation.*

| Investment (USD) | | 23.787.000 |
|---|---|---|
| Operation | | 24 hours/330 days/year |
| Life of project | | 10 years |
| Variable costs | Operation and Maintenance | 8% of fixed capital* |
| | Steam (USD) | 2,746,000 |
| | Power (USD) | 3,846,000 |
| | Chemicals (USD) | 3,751,000 |

**Mechanical pulping is considered an aggressive process.*

**Table 8.**
*Economic analysis parameters for APMP mill.*

| Pulp sale price (USD/ADT) | Biomass maximum purchase price (USD/dry ton) |
|---|---|
| 350 | 15.2 |
| 400 | 58.2 |
| 500 | 144 |

**Table 9.**
*Result of economic analysis for APMP mill.*

Considering utilities equal to zero to calculate the maximum price of biomass and considering the APMP selling price between 350 and 500 USD/ADT, different scenarios could be formulated. For this analysis, the recovery of chemical products or energy that would occur if the plant were related to a kraft mill was not considered. In this case, the profits would be even higher. The results are shown in **Table 9**.

Pulp production seems to be a more attractive business than energy production, although investments in both processes are quite different. However, if the selling price of the pulp falls to 332 USD/ADT, the cost of biomass at the mill gate becomes zero.

## 5. Conclusions

For the final disposal of massive waste generated by the forestry industry, three alternatives were proposed: the generation of energy through burning, the production of filler pulp for paper production, and the production of bioethanol.

According to the experimental results, the production of bioethanol was inhibited in the enzymatic hydrolysis stage, and a more detailed study is required to determine the causes. Thus, APMP pretreatment does not seem to be a viable option for bioethanol manufacturing.

The production of APMP from wood residues is possible and results in good-quality pulp. Therefore, the process can be considered by large kraft pulp production industries for the management of their wood residues.

The production of both power and APMP may be economically feasible. In both cases, the utility of the plant is extremely sensitive to the sale price of energy or pulp. In energy production, sales prices of USD 70/MWh are required, which is normally higher than the cost of wind energy but has been possible in some scenarios. In the production of APMP, higher profits are obtained, but the profits depend strongly on the sale price, and a decrease to less than 332 USD/ADT makes the process unfeasible.

## Acknowledgements

The authors thank the National Agency for Research and Innovation (ANII), which has financed this project through their sectorial energy fund (FSE). They also thank UPM for providing the pin chips with which this work was carried out. The authors particularly thank the Department of Materials and Forest Products of the Technological Laboratory of Uruguay (LATU) for allowing us to execute part of the experimental activity in their facilities.

## Conflict of interest

The authors declare no conflict of interest.

## Author details

Leonardo Clavijo*, Mairan Guigou, Norberto Cassella, Gastón Cortizo, Florencia Risso, Lucía Velazco, Mario Daniel Ferrari, Claudia Lareo and María Noel Cabrera
Universidad de la República, Montevideo, Uruguay

*Address all correspondence to: lclavijo@fing.edu.uy

## References

[1] Korpinen R. On the Potential Utilisation of Sawdust and Wood Chip Screenings. Turku-Finland: Åbo Akademi University; 2010

[2] Fujimoto S, Fukuda T, Kuroda M, Sasaki Y, Sakanishi K, Minowa T, et al. System efficiency and economical analysis of system for producing energy material from wooden biomass. International Energy Journal. 2006;7:289-298

[3] Görgens JF, Carrier M, García-Aparicio MP. In: Seifert T, editor. Bioenergy from Wood. Vol. 26. Dordrecht: Springer Netherlands; 2014

[4] Gerssen-Gondelach SJ, Saygin D, Wicke B, Patel MK, Faaij APC. Competing uses of biomass: Assessment and comparison of the performance of bio-based heat, power, fuels and materials. Renewable and Sustainable Energy Reviews. 2014;**40**:964-998

[5] Integrated Pollution and Prevention Control. Best Available Techniques (BAT) Reference Document for the Production of Pulp, Paper and Board. 2015. Available from: https://eippcb.jrc.ec.europa.eu/sites/default/files/2019-11/PP_revised_BREF_2015.pdf

[6] Bajpai P. Generation of waste in pulps and paper Mills. In: Bajpai P, editor. Management of Pulp and Paper Mill Waste. 1st ed. Springer: Cham; 2015. pp. 9-17

[7] Widell H. Industrial-scale biomass combustion plants: Engineering issues and operation. In: Rosendahl L, editor. Biomass Combustion Science, Technology and Engineering. Woodhead Publishing Limited; 2013. pp. 225-277

[8] Integrated Pollution Prevention and Control. Best Available Techniques (BAT) Reference Document for Large Combustion Plants - Industrial Emissions Directive 2010/75/EU [Internet]. 2017. Available from: https://eippcb.jrc.ec.europa.eu/reference/ [Accessed: July 25, 2022]

[9] Salomón M, Savola T, Martin A, Fogelholm CJ, Fransson T. Small-scale biomass CHP plants in Sweden and Finland. Renewable and Sustainable Energy Reviews. Elsevier Ltd. 2011;**15**:4451-4465

[10] Carvalheiro F, Duarte LC, Gírio FM. Hemicellulose biorefineries: A review on biomass pretreatments. Journal of Scientific & Industrial Research. 2008;**67**(November):849-864

[11] Kim JS, Lee YY, Kim TH. A review on alkaline pretreatment technology for bioconversion of lignocellulosic biomass. Bioresource Technology. 2016;**199**(2016):42-48. DOI: 10.1016/j.biortech.2015.08.085

[12] Kumar AK, Sharma S. Recent updates on different methods of pretreatment of lignocellulosic feedstocks: A review. Bioresources and Bioprocessing. 2017;**4**(1):1-19

[13] Zhao X, Zhang L, Liu D. Biomass recalcitrance. Part II: Fundamentals of different pre-treatments to increase the enzymatic digestibility of lignocellulose Xuebing. Biofuels, Bioproducts and Biorefining. 2012;**6**(5):561-579

[14] Lehto JT, Alén RJ. Chemical pretreatments of wood chips prior to alkaline pulping - a review of pretreatment alternatives, chemical aspects of the resulting liquors, and pulping outcomes. BioResources. 2015;**10**(4):8604-8656

[15] MCS d M, Colodette JL, Jääskelainen AS. Alkaline peroxide mechanical pulping of novel Brazilian eucalyptus hybrids. BioResources. 2012; 7(3):3823-3836

[16] Kruzolek C. Aplicación de variantes del proceso de pulpado al peróxidoalcalino (APMP) a Eucalyptus grandis de 6 y 16 años. [Magíster en Madera, Celulosa y Papel]. Argentina: Universidad Nacional de Misiones; 2000

[17] He W, Hu H, Li W. Multivariate-parameter optimization of the alkaline peroxide mechanical pulp (APMP) process for larch (Larix gmelinii Rupr.) using box-Behnken design. Holzforschung. 2013;**67**(7):727-734

[18] Hu K, Ni Y, Zhou Y, Zou; Xuejun. Substitution of hardwood Kraft with aspen high-yield pulp in lightweight coated wood-free papers part I. synergy of basestock properties. TAPPI Journal. 2006;**5**:21-26

[19] Hespell RB. Extraction and characterization of hemicellulose from the corn fiber produced by corn wet-milling processes. Journal of Agricultural and Food Chemistry. 1998;**8561**(97): 2615-2619

[20] Ebringerova A, Heinze T, Ebringerová A, Heinze T. Xylan and xylan derivatives - biopolymers with valuable properties, 1: Naturally occurring xylans structures, isolation procedures and properties. Macromolecular Rapid Communications. 2000;**21**(9):542-556

[21] Parajó JC, Alonso JL, Vázquez D. Effect of selected operational variables on the susceptibility of NaOH-pretreated pine wood to enzymatic hydrolysis: A mathematical approach. Wood Science and Technology. 1994;**28**(4):297-300

[22] Park YC, Kim JS. Comparison of various alkaline pretreatment methods of lignocellulosic biomass. Energy. 2012; **47**(1):31-35. DOI: 10.1016/j.energy.2012.08.010

[23] Al-dajani WW, Tschirner U. Alkaline extraction of hemicelluloses from aspen chips and its impact on subsequent kraft pulping. In: Engineering, Pulping & Environmental Conference. Tappi; 2007

[24] Lehto J, Alén R. Alkaline pre-treatment of hardwood chips prior to delignification. Journal of Wood Chemistry and Technology. 2013;**33**(2): 77-91

[25] Svärd A, Brännvall E, Edlund U. Rapeseed straw as a renewable source of hemicelluloses: Extraction, characterization and film formation. Carbohydrate Polymers. 2015;**133**: 179-186. DOI: 10.1016/j.carbpol.2015.07.023

[26] Svärd A, Moriana R, Brännvall E, Edlund U. Rapeseed straw biorefinery process. ACS Sustainable Chemistry and Engineering. 2019;7(1):790-801

[27] Wang M, Li Z, Fang X, Wang L, Qu Y. Cellulolytic enzyme production and enzymatic hydrolysis for second-generation bioethanol production. Adv Biochem Eng Biotechnol. 2012;**128**:1-24

[28] Yao M, Wang Z, Wu Z, Qi H. Evaluating kinetics of enzymatic saccharification of lignocellulose by fractal kinetic analysis. Biotechnology and Bioprocess Engineering. 2011;**16**(6): 1240-1247

[29] Rabelo SC, Andrade RR, Maciel Filho R, Costa AC, Filho RM, Costa AC. Alkaline hydrogen peroxide pretreatment, enzymatic hydrolysis and fermentation of sugarcane bagasse to

ethanol. Fuel. 2014;**136**:349-357. DOI: 10.1016/j.fuel.2014.07.033

[30] Dutra ED, Santos FA, Alencar BRA, Reis ALS, de Souza R, FR d, et al. Alkaline hydrogen peroxide pretreatment of lignocellulosic biomass: Status and perspectives. Biomass Conversion and Biorefinery. 2018;**8**(1): 225-234

[31] Liu W, Yuan Z, Mao C, Hou Q, Li K. Removal of hemicelluloses by NaOH pre-extraction from aspen chips prior to mechanical pulping. BioResources. 2011; **6**(3):3469-3480

[32] Latibari AJ, Pourali K, Roghani AF. Alkaline peroxide mechanical pulping of fast-growth paulownia wood. BioResources. 2012;**7**(1):265-274

[33] Li J, Lu M, Guo X, Zhang H, Li Y, Han L. Insights into the improvement of alkaline hydrogen peroxide (AHP) pretreatment on the enzymatic hydrolysis of corn Stover: Chemical and microstructural analyses. Bioresource Technology. 2018;**265**(March):1-7. DOI: 10.1016/j.biortech.2018.05.082

[34] Gupta R, Lee YY. Pretreatment of corn Stover and hybrid poplar by sodium hydroxide and hydrogen peroxide. Biotechnology Progress. 2010;**26**(4): 1180-1186

[35] Rochón E, Cabrera MN, Scutari V, Cagno M, Guibaud A, Martínez S, et al. Co-production of bioethanol and xylosaccharides from steam-exploded eucalyptus sawdust using high solid loads in enzymatic hydrolysis: Effect of alkaline impregnation. Industrial Crops and Products. 2022;**175**

[36] Solier YN, Schnell CN, Cabrera MN, Zanuttini MÁ, Inalbon MC. Alkali-peroxide extraction of xylan from sugar cane bagasse. Characteristics and film forming capacity. Industrial Crops and Products. 2020;**145**(January):112056

[37] Smook GA. Handbook for Pulp and Paper Technologists. 3rd ed. Atlanta, USA: Tappi Pr. 2003. p. 425

[38] Area MC, Kruzolek C. Aplicación de variantes del proceso de pulpado al peróxido alcalino de Eucalyptus grandis de 6 y 15 años. In: Congreso Iberoamericano de Investigación en Celulosa y Papel Ciadicyp 2000. Puerto Iguazú. 2000

[39] Felissia FE, Area MC. Pulpa mecánica al peróxido alcalino (apmp) de Eucalyptus dunii. In: XII Jornadas Técnicas de la Celulosa y el Papel, CONGRESO ATCP. Concepción, Chile: ATCP; 2007

[40] Xu EC. Some of the latest investigations in P-RC APMP pulping of hardwood. Part 1: LCR at secondary refining. In: 88th PAPTAC Annual Meeting. Montreal: PAPTAC; 2002

[41] Zhang H, Huang S, Wei W, Zhang J, Xie J. Investigation of alkaline hydrogen peroxide pretreatment and tween 80 to enhance enzymatic hydrolysis of sugarcane bagasse. Biotechnology for Biofuels. 2019;**12**(1):107

[42] De Carvalho DM, De Queiroz JH, Colodette JL. Assessment of alkaline pretreatment for the production of bioethanol from eucalyptus, sugarcane bagasse and sugarcane straw. Industrial Crops and Products. 2016;**94**:932-941

[43] Alvarez-Vasco C, Zhang X. Alkaline hydrogen peroxide (AHP) pretreatment of softwood: Enhanced enzymatic hydrolysability at low peroxide loadings. Biomass and Bioenergy. 2017;**96**:96-102

[44] Bondesson PM, Dupuy A, Galbe M, Zacchi G. Optimizing ethanol and

methane production from steam-pretreated, phosphoric acid-impregnated corn Stover. Applied Biochemistry and Biotechnology. 2015; **175**(3):1371-1388

[45] Hu F, Jung S, Ragauskas A. Pseudo-lignin formation and its impact on enzymatic hydrolysis. Bioresource Technology. 2012;**117**:7-12

[46] Kumar R, Hu F, Sannigrahi P, Jung S, Ragauskas AJ, Wyman CE. Carbohydrate derived-pseudo-lignin can retard cellulose biological conversion. Biotechnology and Bioengineering. 2013; **110**(3):737-753

[47] Leskinen T, Kelley SS, Argyropoulos DS. E-beam irradiation & steam explosion as biomass pretreatment, and the complex role of lignin in substrate recalcitrance. Biomass and Bioenergy. 2017;**103**:21-28

[48] Rigual V, Santos TM, Domínguez JC, Alonso MV, Oliet M, Rodriguez F. Combining autohydrolysis and ionic liquid microwave treatment to enhance enzymatic hydrolysis of Eucalyptus globulus wood. Bioresource Technology. 2018;**251**:197-203

[49] Sannigrahi P, Kim DH, Jung S, Ragauskas A. Pseudo-lignin and pretreatment chemistry. Energy & Environmental Science. 2011;**4**(4): 1306-1310

[50] Sun S, Cao X, Sun S, Xu F, Song X, Sun RC, et al. Improving the enzymatic hydrolysis of thermo-mechanical fiber from Eucalyptus urophylla by a combination of hydrothermal pretreatment and alkali fractionation. Biotechnology for Biofuels. 2014;**7**(1):116

[51] Pönni R, Galvis L, Vuorinen T. Changes in accessibility of cellulose during Kraft pulping of wood in deuterium oxide. Carbohydrate Polymers. 2014;**101**:792-797

[52] Peters M, Timmerhaus K, West R. Plant Design and Economics for Chemical Engineers. 5th ed. McGraw-Hill Education; 2002. 1008 p

[53] De Garcia Soria X, Villasante C, Cabrera C. Evaluación Económico – Financiera- Proyecto 10 MW Generación de Electricidad a partir de Residuos y/o Subproductos de Biomasa. Montevideo-Uruguay; 2008

[54] Cabrera M, Cocchiararo F, Figares M, Giradello S, Palombo V. Diseño de una planta de fabricación de ceulosa CTMP. Montevideo: Facultad de Ingeniería, Universidad de la República; 2018

Chapter 3

# Nuclear Waste Hazard Reduction

*Hiromichi Fumoto*

## Abstract

This chapter reviews the history of nuclear fuel reprocessing. The implementation of President Carter's International Nuclear Fuel Cycle Evaluation (INFCE) Program hinders the sound fostering of nuclear fuel cycle technologies in the USA and scattered their nuclear engineers to other fields of industries. They once wanted to contribute to developing the "Atoms for Peace Policy." The statement by President Carter changed the nuclear fuel policy as if direct disposal of spent nuclear fuels was quite normal and nuclear fuel reprocessing exceptional. Although the purpose of INFCE is to stop the proliferation of atomic bombs, we experienced and witnessed the proliferation of atomic bombs, despite the banning of nuclear fuel recycling policy for civil purposes. This chapter focuses on "Atoms for Peace" and gives the future perspectives of the nuclear power system at a glance at global ecology. The unnecessary fear of radiation and radioactive substances through the mushroom cloud over Nagasaki and Hiroshima will be discussed as societally important aspects for our future.

**Keywords:** nuclear fuel reprocessing, partitioning and transmutation, geological disposal, radioactive waste, ecology, radiation protection

## 1. Introduction

This chapter reviews the history of nuclear power development for civil purposes in the context of nuclear fuel and radioactive substances. Soon after the Atoms for Peace speech addressed by Mr. Dwight D. Eisenhower, President of the United States of America, to the 470th Plenary Meeting of the United Nations General Assembly [1], confidential documents were declassified and delivered to the nonnuclear weapon states to have world share the technology of nuclear power for peaceful uses. The United States of America had led the development of nuclear power to develop and construct light water reactor (LWR), witnessing a sharp rise of a nuclear share in the power grid. At the same time, following the nuclear power growth, nuclear fuel supply was becoming one of the key interests. Recycling uranium and plutonium had been envisaged to be the most feasible at this stage. The nuclear fuel reprocessing plant for civil purposes was planned and constructed, first in West atomic Valley in New York State in 1966 [2] and second in Barnwell in South Carolina State in 1970 [3]. The diagram of process flow in those plants was similar to that of plants for military purposes. However, their plutonium product will be recycled and mixed with uranium as mixed oxide fuel that uranium is planned to be transferred to the $UF_6$ conversion plant and fed to the uranium enrichment plant [4].



IntechOpen

The typical concept of nuclear fuel cycle is shown in **Figure 1** [5].

As is described in **Figure 1**, spent fuel reprocessing plants were once constructed and operated in the U.S. but are now omitted. This chapter discusses the history and thinks about where the rationale is.

Reprocessing plant is a key component of the nuclear fuel cycle to recover uranium and plutonium, and nuclear fissile materials are to be used for nuclear fission, so that we can recycle nuclear fuels as much as possible. It must be an environmentally friendly means of saving nuclear fuel consumption to lessen its impact on the earth through human activities. It contributes to extracting less amount of uranium through mining for a sustainable society.

Nevertheless, the nuclear fuel recycling policy had to be ended in the USA. Nuclear Non-Proliferation Act of 1978 Statement is signed into law [6]. It refers to the Atomic Energy Act of 1954 and the adoption of the Non-Proliferation Treaty by the United Nations in 1968. The statement shows disagreement with the unnecessary commitment to the commercialized use of plutonium, which may not be able to prevent the proliferation of nuclear weapons. Thus, nuclear fuel reprocessing was omitted and the so-called once-through fuel cycle, deleting the fuel reprocessing, and recycling illustrated in **Figure 1**, was adopted as the U.S.A. policy of nuclear energy utilization.

Current nuclear fuel cycle in the U.S. is shown below [7].

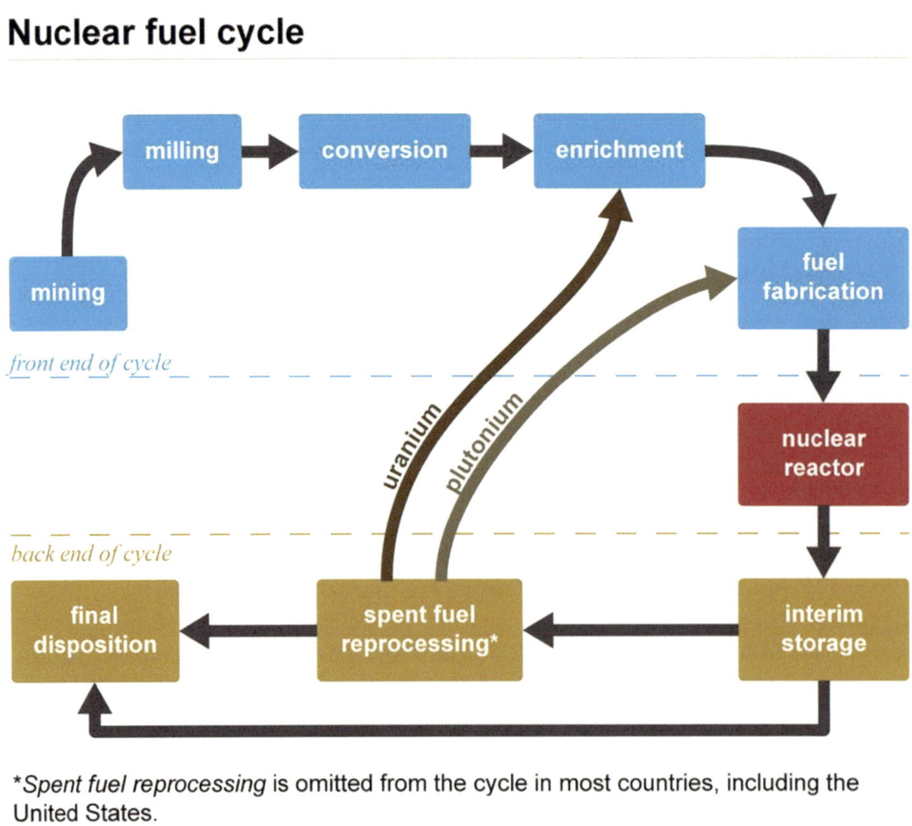


**Figure 1.**
*Concept of nuclear fuel cycle [5].*

Spent fuel reprocessing is not shown in **Figure 2** but remarked in used fuel as "pending possible reprocessing or permanent storage." This note is the reflection of the statement of the Nuclear Non-Proliferation Act of 1978, which states that "I continue to oppose making premature and unnecessary commitments to commercialization of the fast breeder reactor and reprocessing, as exemplified in the United States by the Clinch river and Barnwell projects" [6]. The Clinch River project is the development of the fast breeder reactor, which consumes and produces plutonium simultaneously. The Barnwell projects are the development of the commercial spent fuel reprocessing plant and its related facilities. The facilities at the planning stage are a uranium conversion plant for recovered uranium and a mixed oxide (MOX, a mixture of uranium oxides and plutonium oxides) fuel plant for the recovered plutonium planned to be built in the adjacent area.

The statement of President Jimmy Carter states that "we are premature to commercialize the utilization of plutonium as an energy source, so it reflects in **Figure 2** as noted if we become mature enough to get rid of nuclear weapon proliferation risks, it may be turned to be "possible reprocessing," otherwise, "permanent storage" is the only one option [6].

It states further that "the U.S.A. withdrawn all the program of commercial utilization of plutonium, however, it also asks the international community to abandon their promotion of plutonium and weapon-grade enriched uranium for commercial purposes." It extends the setup of INFCE, International Nuclear Fuel Cycle Evaluation with more than 40 nations is now going. One of the agreements reached by INFCE is that uranium enrichment is limited to up to 20% maximum for the civilian reactors. This mainly targets test reactors that use almost weapon-grade enriched uranium to replace the 93% enriched uranium with 20% enriched uranium. Not limited to research reactors, INFCE ruined Germany's high-temperature reactor development. The German development policy for high-temperature reactor development and how the INFCE affects them will be discussed below.

In 1961, the German prototype reactor of AVR started its construction [8]. Its nuclear fuel is the so-called pebble bed design fuel. Still, this fuel type is one of the

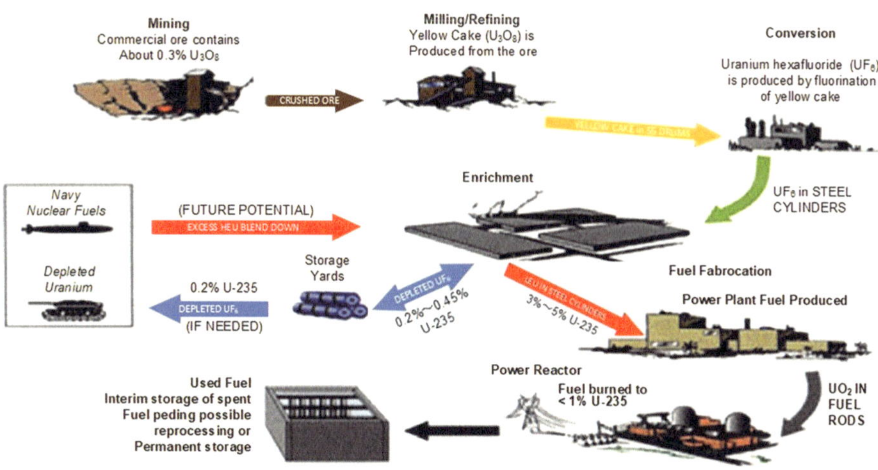


**Figure 2.**
*Current nuclear fuel cycle in the U.S.A. [7].*

candidate fuel designs expected to be adopted in our next-generation reactors. AVR is a helium-cooled high-temperature prototype reactor and foresees the future utilization of nuclear energy even for iron manufacturing. Its nuclear fuel cycle shall be focused on INFCE and its influence on fuel cycle development. Originally, the drive fuel of AVR was highly enriched, around 93% enriched uranium. The core is surrounded by thorium fuel, and the neutron irradiation breeds $^{233}U$, which can be used as fissile material to replace highly enriched uranium used in its initial core.

**Figure 3** shows the $^{233}U$ recovery process from the irradiated thorium fuel [9]. The $^{233}U$ produced in the irradiated thorium and the thorium as matrix materials are dissolved in highly concentrated nitric acid and extracted together with tri-butyl phosphate (TBP). Most of the fission products, including protactinium, are separated, leaving in the aqueous phase through the liquid–liquid two phases (aqueous and solvent, TBP with diluent) extraction. Then, the extracted thorium and $^{233}U$ are stripped to the aqueous phase separately, using the difference in their distribution ratios between the aqueous to solvent phases. Since protactinium decays to $^{233}U$, the recovery of protactinium is also proposed in the effluent treatment.

The recovered $^{233}U$ will be recycled for fuel fabrication, as in the plutonium case in **Figure 1**. It is widely known that $^{239}Pu$ and $^{235}U$ are fissile materials used for nuclear weapons, although $^{233}U$ is not as famous as them. Since it was only produced through irradiated thorium fuel reprocessing, Germany is the only state that tried to recover $^{233}U$ through reprocessing, using a new type of reactor on an industrial scale. INFCE set the criteria of 20% enriched $^{235}U$ as the upper limit of enrichment for civil purposes. Thus, the research reactors were forced to redesign their core configuration to lower the enrichment down to less than 20%. For $^{233}U$, INFCE put the threshold of 12% diluted by $^{238}U$ [10]. In the original perspective of the thorium fuel cycle, the incentive is to obtain $^{233}U$, which is not contaminated with other uranium isotopes. However, INFCE had ordered to dilute the recovered $^{233}U$ with $^{238}U$, which ruined the feasibility of the thorium fuel cycle in Germany.

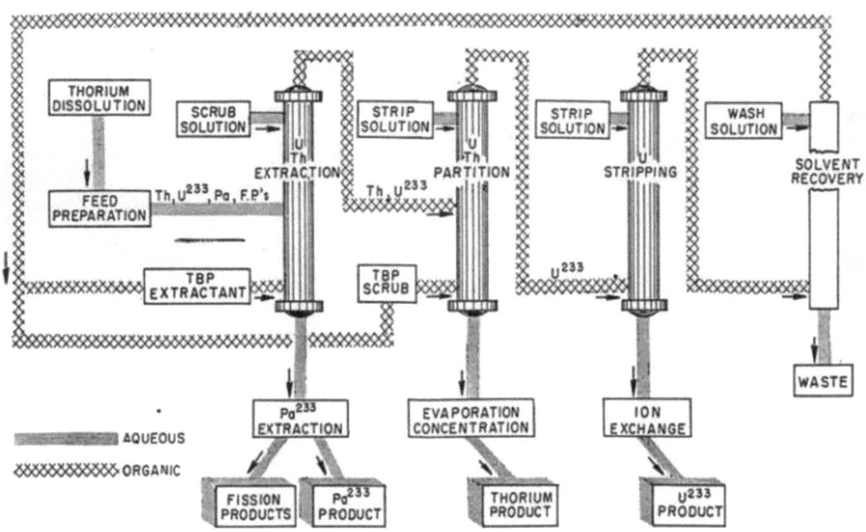


**Figure 3.**
*Schematic diagram of $^{233}U$ recovery process [9] (courtesy of International Atomic Energy Agency, copyright IAEA).*

As discussed before, the statement of President Carter on nuclear non-proliferation has not only abandoned nuclear fissile material recycling in the U.S.A. but also limits the fissile materials reuse internationally. The statement named France and Japan as those states to choose the policy of adopting the nuclear fuel cycle and are still keeping the policy of recycling plutonium for civil purposes [6].

## 2. Long-lived radioactive nuclides produced by nuclear power generation

Nuclear energy is delivered when the fissile material is fragmented into fission products. They are radioactive. Sometimes, the utilization of nuclear energy is criticized because they produce human-made radioactivity. However, fossil fuels have also been producing carbon dioxides and the author believes that to what extent we can afford to accept them must be important. Most of the half-lives of fission products are relatively short, and their radioactivity decreases quickly with time. **Figure 4** shows the typical decrease of radioactivity in spent fuel [11].

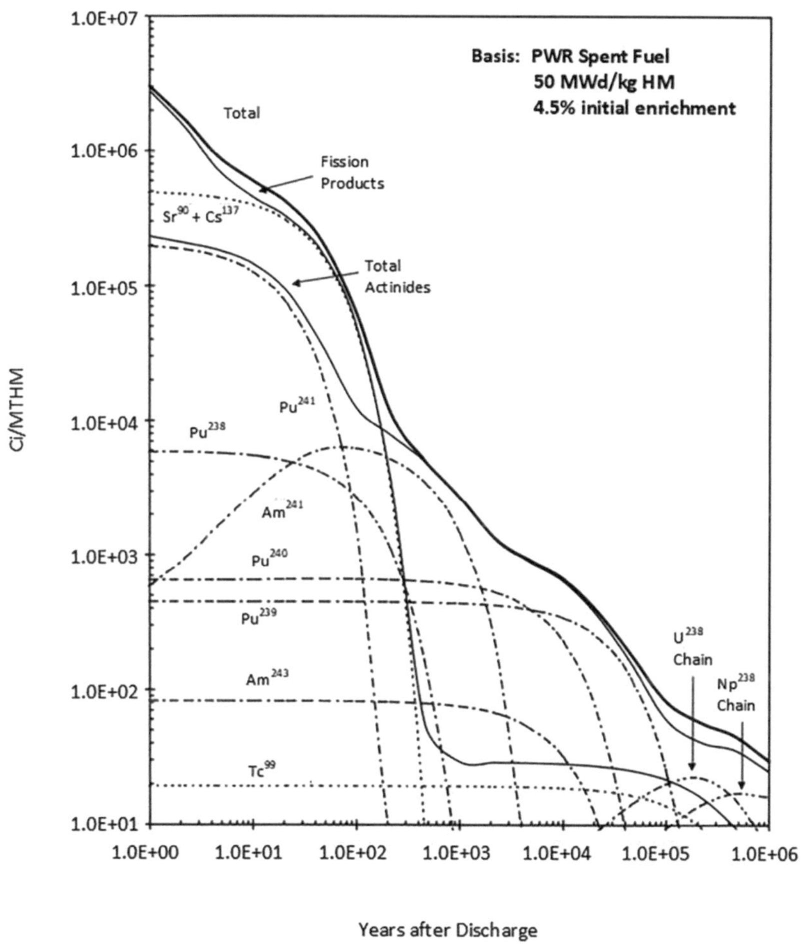


**Figure 4.**
*Radioactivity decreases with time (in the case of PWR fuel with burn up of 50 MWd/kg(Copy of source captions) and initial enrichment of 4.5%) [11].*

**Figure 4** uses the unit of Ci/MTHM, and curies are easier for us to grasp the magnitude of radioactivity. The unit Curie is named following her honorable achievement of purifying radium from thousands of tons of ores from the mines of St Joachimsthal, Germany (today, the mines of Jáchymov in Check) [12]. The unit of a Curie is defined as radioactivity in 1 g radium. In **Figure 5**, radioactivity starts with 3 x $10^6$ curies, equivalent to 3 tons of radium in radioactivity from each 1 ton of nuclear fuel. The radioactivity falls to be equivalent to 3 g radium when one million years have passed. If we believe Mrs. Curie collected 1 g of radium, we may accept 3 g of radium equivalent in our world. However, if 3 tons of radium exist in our vicinities, the residents may become awful to abandon the generation of radioactive waste immediately. It is a typical sentiment but acceptable since we witnessed the atomic bomb cloud over the peaceful city of Hiroshima and Nagasaki in 1945.

However, the author believes scientists must deliver the rationale for assessing radioactivity hazards. As discussed before, the world of nuclear energy is divided into

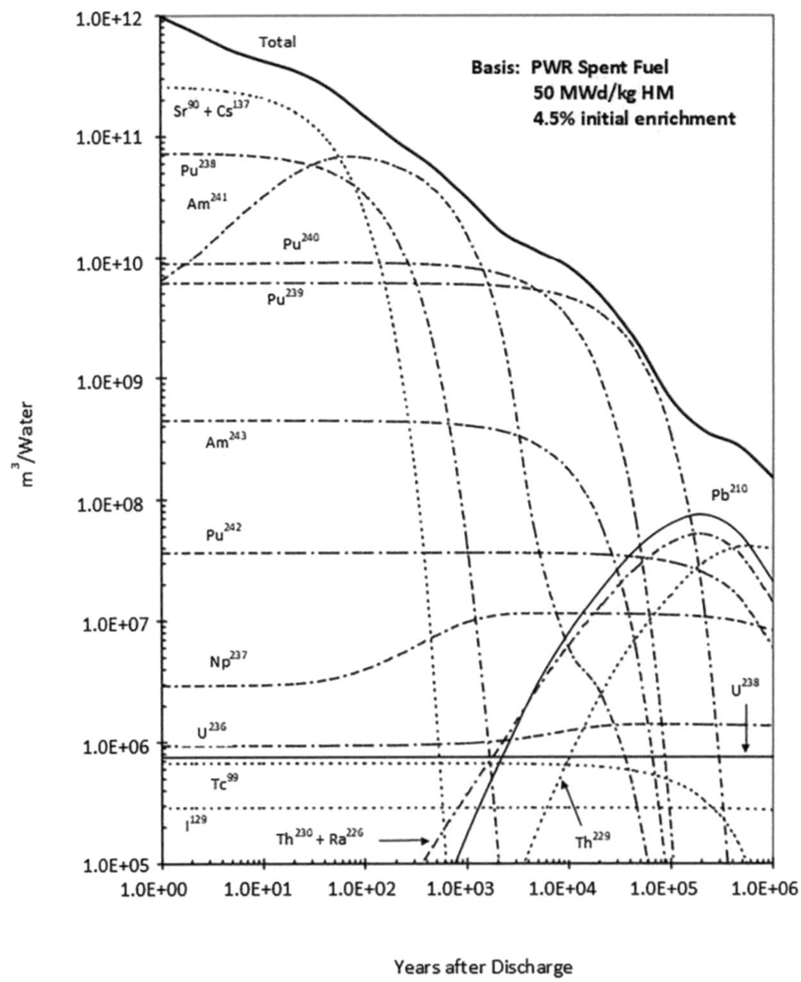


**Figure 5.**
*Radiotoxicity and time after discharge, radiotoxicity shown as dilution water volume to be accepted as not hazardous [11].*

two major groups in terms of nuclear fuel cycles. One is the so-called once-through cycle, which does not process the spent fuel; thus, the radioactivity it generates is left as it is. In this case, the radioactivity stored or disposed of is the same as in **Figure 4**.

On the other hand, if we reprocess the spent fuel, the activity of plutonium and uranium would be -thousandth of the values shown in **Figure 5**. As shown in this figure, fission products generated through nuclear power are relatively quickly down to 3 g of radium in around a few hundred to one thousand years. After the decay of most fission products, the major contributors to the residual radioactivity are actinide elements. Thus, if we wish to reduce the radioactivity to a few grams of radium within one thousand years, the actinide elements, such as americium and curium, must be recovered in addition to plutonium and uranium. Those elements are recovered through the existing reprocessing plant in the PUREX process. The details of actinide recovery (among nuclear technology experts, the term minor actinide is widely used to identify actinide elements except for plutonium and uranium, focusing on the elements not recovered by PUREX reprocessing plant) are discussed later. However, further discussion shall be made here if the once-through fuel cycle is viable or not in its nature in ecology.

The radioactivity of each nuclide does not directly indicate the radiotoxicity to human health. The higher the energy irradiated from the radioactive substances is, the more severe their consequences are to our body showing radiation detriment. Radiotoxicity is normally expressed as the amount of water needed to dilute the radionuclide concentration to tolerably acceptable levels. As indicators of acceptable levels, annual levels of intake (ALIs) are used [13, 14]. ALIs are levels of specific radioactivity concentration acceptable to our daily lives. ALIs are evaluated to be equivalent to our exposure received by the radionuclides to be 1 mSv/year for the general public. The 1 mSv/year is expressed as the dose limit in radiation protection practices agreed upon with International Committee on Radiation Protection (ICRP) [14].

We receive 2.4 mSv/year averaged over the population in our globe [15]. Thus, it would be strange if we could keep the code of dose limit of 1 mSv/year, while we exceed the limit in living every day. To manage the discrepancy, ICRP prepared a note of implementation in 1 mSv/year, when they first introduced the dose limit concept, saying that the level of 1 mSv/year was applied to the cases in controlling radiation exposure and not to be applied to those cases in an uncontrollable situation. In other words, 1 mSv/year is not a bounding limit for judging if the environment is safe or not [16].

**Figure 5** shows an example of radiotoxicity interpreted from the radioactivity as shown in **Figure 4**. The volume of water needed to dilute the radioactivity into the exposure levels of 50 mSv/year is shown as radiotoxicity. The exposure limit adopted for waste disposal is one of the key societal issues since the regulators support the principle of the hypothesis that there are no threshold levels of radiation exposure in terms of their effects on human health. In most cases, the general public believes that radiation is as low as possible. However, the concept of no threshold can also be applied to an idea of acceptable radiation to be decided by the society or community, namely, to which extent the additional man-made radionuclides may be accepted in the community even though they generate a slight rise of environmental radiation levels near the disposal site.

**Figure 5** indicates that most of the fission products represented by Cs and Sr. are to be decayed in several hundreds of years, and most residual activities come from actinides. Suppose we shorten the time frame of the isolation period of radioactivity from one million years to a few hundred years. In that case, the sentiment of fear of radioactive waste might be eliminated. It will be discussed in later sections.

## 3. Partitioning and transmutation of minor actinides

Spent fuel reprocessing plant using PUREX process partitions uranium and plutonium to recycle them to nuclear power plants. The solvent of TBP was found to be suitable for those two elements to separate them from other fission products. As shown in **Figure 5**, to reduce the radioactivity over the extended period of over 1000 years, residual actinides, the so-called minor actinides must be separated from fission products. If the reprocessing plant could be regarded as an ecologically friendly plant, it must be the due course of our destined R&D step further to recover minor actinides.

**Figure 6** shows the contribution of separating uranium and plutonium by spent fuel reprocessing and separating minor actinide, respectively [17].

In **Figure 5**, radiotoxicity is indicated by $m^3$ water to dilute the radioactivity to give the exposure levels of 50 mSv/year, while **Figure 6** shows the radiotoxicity by Sv/tsm. They are indicating the radiotoxicity, the same but in a different index. **Figure 5** shows the amount of water diluted to reach a certain level of exposure to the radionuclides. If not diluted, the exposures to be taken are to be multiplied by the volume of water for dilution. Even though each researcher defines the radiotoxicity scales differently, choosing the radiotoxicity levels to be targeted must be mutually agreed upon among researchers. The target levels are not the dose limit given by ICRP [14], exemption or clearance levels given by IAEA [18], but the levels observed in our surrounding environment. In **Figure 6**, the natural uranium levels are shown as target of radiotoxicity in time of isolation from our biosphere. **Figure 6** indicates that the isolation period for radioactive waste resulting in nuclear power generation can be dramatically shortened from 170,000 years to 330 years. Thus, researchers' challenge to remove minor actinides has never been given up in the past, now, and in the future.

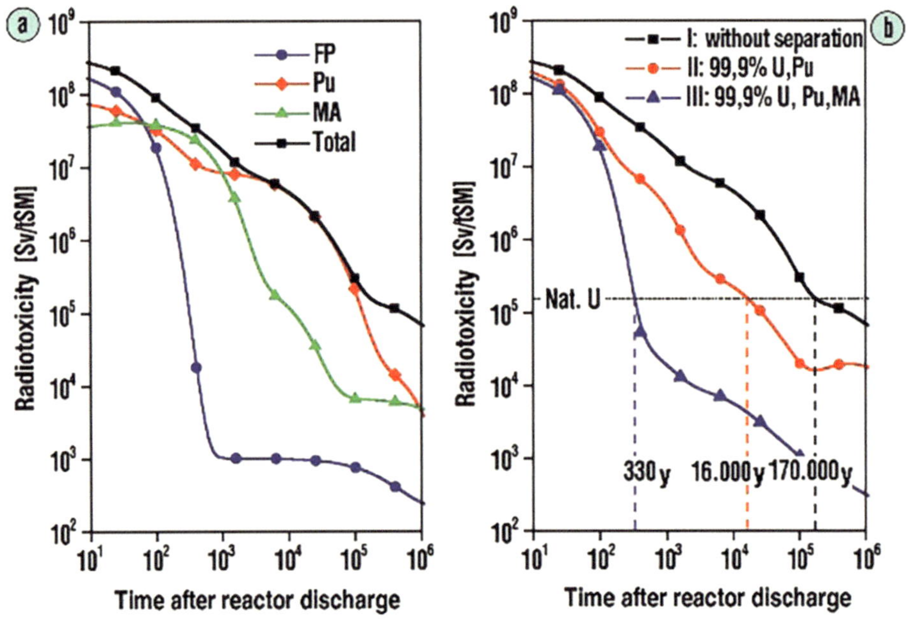


**Figure 6.**
*The contribution of reprocessing and minor actinide partitioning on radiotoxicity [17].*

The major elements of minor actinides are americium and curium. In the U.S.A, "octyl(phenyl)-N,N-diisobutyl Carbamoyl Methyl Phosphine Oxide" (CMPO) is selected to investigate the recovery of americium from the raffinate of the extraction cycle in PUREX process in 1980s [19]. The process is named as TRUEX process. Since then, many other solvents have been proposed, but the investigation has continued to industrialize the partitioning of minor actinides to the end. In addition to the recovery process of minor actinides, it must be noted that transmuting the minor actinides is also facing obstacles. If we recycle those minor actinides into light water reactor (LWR, widely adopted as a standard nuclear power plant globally), they absorb thermal neutrons again to be transformed into higher atomic number actinides. They may not be able to transform them into shorter half-life nuclides.

To transform the minor actinides into nuclear reactors, we need a reactor using fast neutrons, the so-called fast reactors (FRs). FR used to be called fast breeder reactor (FBR) when the concept of breeding plutonium was accepted. Irradiated by fast neutrons, $^{238}U$ can be transformed into $^{239}Pu$, which is fissionable. Since INFCE, the $^{239}Pu$ breeding feature of fast reactors is becoming gradually unpopular, and the word "breeding" is omitted now [19].

Several other reactors, such as molten salt reactors, had been discussed for minor actinides transmutation [19]. However, in this chapter, FR is adopted as the representing type of reactor to transmute minor actinides. Partitioning and transmutation of minor actinides are not yet industrialized. However, it is worth continuing to work for it since it might be a key to accepting the radioactive waste generated through nuclear power through our highest ethical value to give the lowest damage to the globe.

## 4. How safe is the safe for radiation protection

As concluded by the Ethical Committee in Germany, most people believe that we have not yet been wise enough to give a solution to radioactive waste [20]. The committee's report says that if we cannot get the expected probability value, we must judge if it is acceptable, assessing the worst-case consequences. What would be the worst case for radioactive waste? Since we cannot imagine our world 170,000 years later, far in the future, let someone believes that the worst case might be a virtual scenario of the waste coming up to the surface of our biosphere in 10,000 years.

From **Figure 6**, direct disposal is 10 to $10^2$ times higher than natural uranium in once-through spent fuel as radioactive waste. In contrast, reprocessing is a few times higher, and reprocessing plus minor actinides removal is 10 to $10^2$ times lower. Those scientists supporting the R&D for minor actinides removal take it for granted to use the radiotoxicity of natural uranium as reference levels to judge whether the waste's radioactivity is safe to leave the waste as it is in our biosphere environment. A similar approach to our interpretation in uranium mining, extraction, and enrichment is shown in **Figure 7** [21].

**Figure 7** shows the radioactivity of natural uranium had never been changed through our activities up to the loading of nuclear fuel to nuclear power plant but relocating the extracted uranium and uranium progenies (daughters in **Figure 8**), and the extracted uranium is divided into the enriched uranium and depleted uranium. It is clear in the scientific background that chemical or physical processing does not cause any nuclear reaction. Thus, the nuclides supplied to the chemical/physical process cannot be more radioactive than before.

Most scientists might not be interested in **Figure 7** because it is clear. However, the author is keen on the scale of elapsed time to be 10 million years. In scientific meaning, 10 million years is ten times one million years, but in societal meaning, it implies our extension of an unbelievable period. No one can imagine how our society would be. Nevertheless, the radioactivity survived as it is now.

**Figure 7** indicates two facts; one is that the radioactivity of natural uranium is not to be decreased even after ten million years, and the other is that our activity of uranium mining, extraction, and enrichment does not influence our radioactivity for a total of natural background. The former gives the idea that radiation could not be gotten rid of in this globe and the latter gives the idea that uranium extraction and enrichment do not affect radioactivity on earth. The author believes that **Figure 8** enhances the understanding of radiation in the natural environment.

Once we understand the illustration in **Figure 7**, acceptance of radioactive waste disposal becomes easier. **Figure 8** is another example of **Figure 6** but uses radioactivity left behind compared to natural uranium [21].

Some may criticize that radiotoxicity must be calculated, and a simple comparison of radioactivity has no meaning. It is true to scale the effect of radiation shall be measured with the dose of exposure. However, **Figure 8** suggests the ethical values of our common world that the radioactivity generated by energy generation can be remediated to the same levels in the initial stages. Radiation protection controls radiation dose exposure to men/women in its narrower sense. However, once natural radioactivity is included, the concept of exemptions or existing exposures has been introduced to balance the flexibility of radiation protection philosophically. Thus, it is inevitable to introduce some indicators other than exposure doses.

This kind of idea in choosing indicators other than doses always invites rigorous discussions among radiation protection experts. Some radiation protection experts always emphasize that the risk must be measured *via* doses caused by the radiation sources to assess the detriment to our health. They are turning a deaf ear to the assessment of radiotoxicity by cumulative radioactivity or specific radioactivity alone or related parameters, implacably hating the attitude of not mentioning the doses caused by the radioactivity.

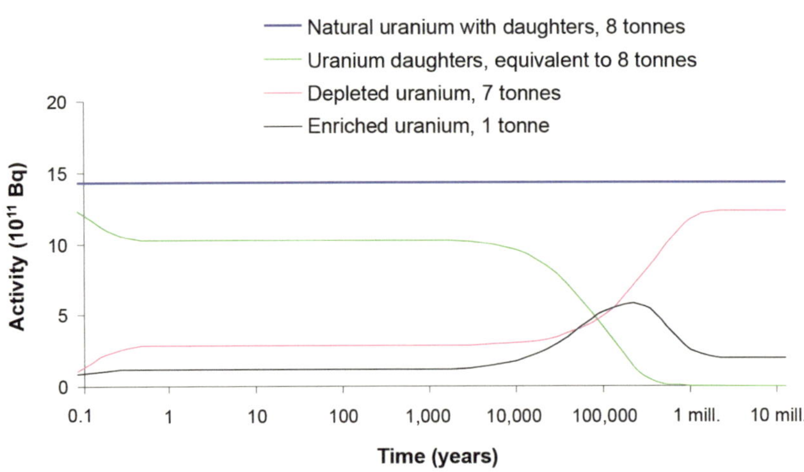


**Figure 7.**
*Development of activities in different fractions of uranium in uranium supply to nuclear power plants (assuming that the 8-ton natural uranium equilibrated with daughters are processed into 7 tons of depleted uranium, 1 ton of enriched uranium) [21] (through the courtesy of Svensk Kärnbränslehantering AB).*

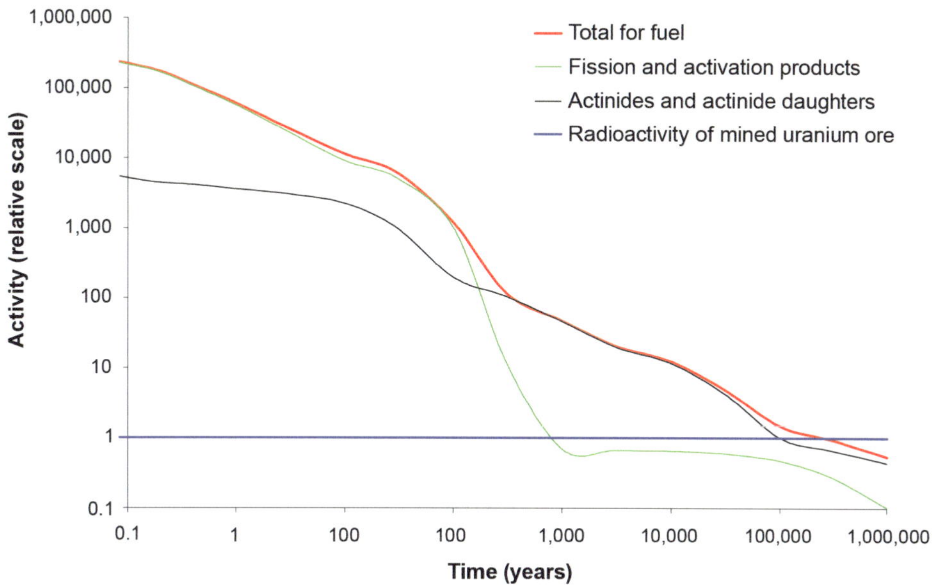


**Figure 8.**
*Relative activity of spent nuclear fuel of type SVEA 64 with a burnup of 38 MWd/kgU(Copy of source captions) [21] (through the courtesy of Svensk Kärnbränslehantering AB).*

We must unify those arguments among scientists to identify the ideologies behind the struggles. They are due to the history of the inventions of radiation applications. In our immature age, we do have much misuse of radium and X-rays in our medical and aviation industries (radium painting for airplane cockpit displays). We witnessed a series of injured patients, X-ray technicians, and radium painters. In the 1930s, the first trial of protecting human rights was completed. ICRP had proposed radiation protection standards and guidelines for protecting radium ingestions [12]. Nevertheless, the world has aligned the radiation applied to the most powerful weapon in killing people. At the same time, research on radiation protection had been enhanced to protect workers of the alliance who developed and produced the atomic bomb [22].

The orchestration of the promotion of atomic bombs to kill the enemy as much as possible and the arrangement of radiation protection to lower the exposures of the alliance side attributes the conclusions to be that radiation is harmful even a bit of exposure to our body. Our hostile attitudes toward the tiny dose of exposure, avoiding any number of milli- or micro-Sieverts could be regarded as the saga of our history in atomic bomb development.

The linear no-threshold (LNT) hypothesis is widely believed in our community, especially among media people. It can be interpreted as radiation exposure being harmful indifferent of the levels of exposure. Even if the exposure is so small, no safety allowance exists for being irradiated. Major experts in radiation protection believe that LNT hypothesis is only a hypothesis and not practically applicable to our society [23]. However, the media values the sensation of radioactive substances as a source of readers' fear, hardening the spell of the LNT hypothesis. They explained that the scientists have not yet agreed with concluding the denial of the applicability of the LNT hypothesis to our societies. In their sense, until the last scientist who supports the LNT application to our daily life of radiation exposure and radioactive substances intake becomes silent, radioactive waste is the risk of cancer in our society even if its exposure is so small.

It has taken less than 10 years to develop an atomic bomb since the discovery of fission in uranium. At the same time, a rational understanding of harmful radiation to our body seems to take more than a century. The latter is needed for the implementation of atoms for a peace accord. The scientists who can give rationales to the background of radiation protection in history are the only contributors to disseminating the meaning of nuclear energy in our society.

## 5. Future perspectives

The author believes nuclear energy plays an important role in saving the globe. Dr. Takashi Nagai, who submitted the first rescue report of Nagasaki atomic bomb victims, mentioned in his "Atomic Bomb Rescue and Relieve Report" that peaceful uses of atomic energy are the only one offering dedicated to the souls lost in the atomic bomb explosion in Japan [24]. The author introduces the countermeasures for reducing radiotoxicity in radioactive waste generated through our nuclear power generation, although they are still in the stage of R&D.

The once-through cycle option is selected in the U.S.A. for political reasons. The Fast Breeder Reactor cannot be in line with the nuclear power plant, since banning plutonium utilization for civil purposes has not yet been lifted. No one can predict when we can freely discuss the utilization of plutonium in our energy production. Without perspectives of fast neutron resources, the separated minor actinides may not be transmuted on an industrial scale.

Our common goal is a sustainable energy supply and nuclear energy is one of the vital candidate energy resources. Radioactive waste is believed to be the last but invulnerable to stop nuclear. We have to recall the history of nuclear energy development, and the scientists with full knowledge of radiation protection in nature will open the gate to receive the acceptance of radioactive waste for society to use nuclear energy in the end.

It must be a long way, tens of or hundreds of years, to accept nuclear energy, but the author believes in the rationales of radiation protection in wider arrangements to our environment, not limited to the narrower arrangement of controlling exposure of doses. The day will come when our society is not too sensitive to radiation as of today, but to accept radiation exposure as it is.

## 6. Conclusions

It has been discussed how to tame nuclear energy to achieve sustainable development goals (SDGs). Some understand it as one of the vital tools to be chosen, while others refute it. Irrational fear of extra radiation exposure to our health can only be rectified by the power of natural and social science studies for our society to understand the history of nuclear energy both for military and civil purposes. That is the only way to understand why nuclear fission is discovered, plutonium production and separation are invented by human beings.

## Author details

Hiromichi Fumoto
Japan Inspection Co., Ltd., Tokyo, Japan

*Address all correspondence to: h.fumoto@nihonkensa.co.jp

## References

[1] Atoms for Peace Speech. Address by Mr. Dwight D. Eisenhower, President of the United States of America, to the 470th Plenary Meeting of the United Nations General Assembly, Tuesday, 8 December 1953, 2:45 p.m. [Internet]. Available from: https://www.iaea.org/about/history/atoms-for-peace-speech

[2] Fuel Reprocessing History. New York State [Internet]. Available from: https://www.nyserda.ny.gov/Researchers-and-Policymakers/West-Valley/Fuel-Reprocessing-History

[3] Croff AG, Wymer RG, Tavlarides LT, Flack JH, Larson HJ. Background, Status, and Issues Related to the Regulation of Advanced Spent Nuclear Fuel Recycle Facilities. Washington, DC: ACNW&M White Paper, NUREG-1909; 2008. Available from: https://www.nrc.gov/docs/ML0815/ML081550505.pdf

[4] Final Safety Analysis Report. Barnwell Nuclear Fuel Plant Separation Facility", Docket 50-332-41. US government report. Allied-Gulf Nuclear Services. Washington, DC. 1973

[5] U.S. Energy Information Administration (EIA). Nuclear explained the nuclear fuel cycle [Internet]. 2022. Available from: https://www.eia.gov/energyexplained/nuclear/the-nuclear-fuel-cycle.php

[6] Jimmy Carter. "Nuclear Non-Proliferation Act of 1978 Statement on Signing H.R. 8638 Into Law" [Internet]. 1978. Available from: https://www.presidency.ucsb.edu/documents/nuclear-non-proliferation-act-1978-statement-signing-hr-8638-into-law

[7] Office of Nuclear Energy. Nuclear Fuel Cycle [Internet]. 2022. Available from: https://www.energy.gov/ne/fuel-cycle-technologies/uranium-management-and-policy/nuclear-fuel-cycle

[8] "AVR Juelich, Germany", The World Nuclear Association [Internet]. Available from: https://www.world-nuclear.org/reactor/default.aspx/AVR%20JUELICH

[9] IAEA-TECDOC-1450. Thorium fuel cycle — Potential, benefits and challenges [Internet]. 2005. Available from: https://www-pub.iaea.org/mtcd/publications/pdf/te_1450_web.pdf

[10] Forsberg CW, Hopper CM, Richter JL, Vantine HC. Definition of Weapons-Usable Uranium-233, ORNL/TM-13517 [Internet]. 1998. Available from: https://thoriumenergyalliance.com/wp-content/uploads/2020/02/weapons-usable-u-233-ORNL-TM-13517.pdf

[11] MIT. The Future of Nuclear Power, an Interdisciplinary MIT Study. Massachusetts, USA: Massachusetts Institute of Technology; 2003

[12] Fumoto H. Radioactive waste disposal (III) -exemption and clearance in waste disposal. Radioisotopes. 2019;**68**(11):773-789

[13] ICRP. 1990 Recommendations of the International Commission on Radiological Protection. Users ed. ICRP Publication 60. Oxford: Pergamon Press; 1991

[14] ICRP. Principles for the disposal of solid radioactive waste. Principles for the Disposal of Solid Radioactive Waste. Recommended citation ICRP. ICRP Publication 46. Annals of ICRP. Oxford: Pergamon Press; 1985;**15**(4). Available from: https://www.icrp.org/publication.asp?id=ICRP%20Publication%2046

[15] UNSCEAR. UNSCEAR 2000 Report Volume I, Sources of Ionizing Radiation. Report to the General Assembly With Scientific Annexes. New York, NY: UN Publications; 2000

[16] ICRP. Recommendations of the ICRP. Available from: https://www.icrp.org/publication.asp?id=ICRP%20Publication%2026ICRP. Publication 26. Recommended citation ICRP. ICRP Publication 26. Annals of ICRP. Oxford: Pergamon Press; 1977;**1**(3)

[17] Freiesleben H. Final Disposal of Radioactive Waste. Vol. 54. EPJ Web of Conference. Les Ulis Cedex A, France: EDP Sciences; 2013. p. 01006. DOI: 10.1051/epjconf/20135401006

[18] IAEA. Radiation Protection and Safety of Radiation Sources: International Basic Safety Standards, General Safety Requirements Part 3. Vienna: IAEA; 2014

[19] IAEA. Assessment of Partitioning Processes for Transmutation of Actinides, TECDOC-1648. Vienna: IAEA; 2010

[20] Ethics Committee for a Safe Energy Supply. Germany's Energy Transition – A Collective Project for the Future [Internet]. 2011. Available from: https://archiv.bundesregierung.de/resource/blob/656922/457334/784356871e5375b8bd74ba2a0bde22bf/2011-05-30-abschlussbericht-ethikkommission-en-data.pdf?download=1

[21] Hedin A. Spent Nuclear Fuel - How Dangerous Is it? A Report from the Project "Description of Risk", Technical Report 97-13. Stockholm, Sweden: Swedish Nuclear Fuel and Waste Management Co; 1997

[22] Fumoto H. Radioactive waste, the last battlefield of science, ethics and law. Journal of Quality in Health care & Economics. 2021;**4**(5):000240. Available from: https://www.medwinpublishers.com/JQHE/radioactive-waste-the-last-battlefield-of-science-ethics-and-law.pdf. DOI: 10.23880/jqhe-16000240

[23] Fumoto H. Ethical Values in Radiation Protection. In: Bioethics in Medicine and Society. London: InTechOpen; 2020. DOI: 10.5772/intechopen.93786

[24] Nagai T. "Atomic Bomb Rescue and Relieve Report", Division of Scientific Data Registry, Atomic Bomb Deceases Institute. Nagasaki: Nagasaki University; 1945 (In Japanese)

Chapter 4

# Revolutionary Plastic Mechanical Recycling Process: Regeneration of Mechanical Properties and Lamellar Structures

*Patchiya Phanthong and Shigeru Yao*

## Abstract

Plastic recycling is one method that can reduce the amount of waste plastics in the environment. Especially in the mechanical recycling process, the waste plastics are reprocessed by mechanical approaches and reproduced as recycled products. Based on the life cycle assessment studies, mechanical recycling shows lower emissions than other approaches which is the most suitable method for environments. However, the mechanical properties of recycled products are much more degraded than virgin plastics. In this chapter, revolutionary plastic mechanical recycling is introduced. The new type of twin-screw extruder with the addition of molten resin reservoir unit is applied with various extrusion conditions. As a result, the mechanical properties of recycled products were recovered than its original materials. The relationship between recycling process condition, mechanical properties, and mesoscale lamellar structure are also discussed. It can be found that steady shear which is a conventional process affected the degradation of mechanical properties. In another way, re-extrusion by using a new type of twin-screw extrusion with a suitable processing condition is related to the regeneration of mechanical properties and lamellar structures similar to virgin plastics. This chapter is expected to introduce the recent advances in plastic mechanical recycling process and propose the future prospects of plastic recycling technology.

**Keywords:** plastic mechanical recycling, mechanical properties, lamellar structure, physical regeneration, waste plastics, recycled plastics, extrusion, twin-screw extruder

## 1. Introduction

### 1.1 Plastic mechanical recycling process

The consumption of plastic has greatly increased all over the world due to being used in daily life. Single-use plastics have become the cause of environmental problems because they were disposed of after one-time use and gathered on land and in the marine environment [1]. The degradation of plastics in the environment takes a very long time, that is, decades to thousands of years [2]. For this reason, a suitable



IntechOpen

way for decreasing the amount of waste plastics is an attractive research subject in order to protect the future world environment.

Recycling is one of the processes that are able to decrease the amount of waste plastics. There were three main categories for plastic recycling; mechanical recycling, feedstock recycling, and energy recovery [3, 4]. Among them, mechanical recycling was attractive because it was evaluated to have a high environmental performance from the life cycle assessment (LCA) evaluation. The use of a mechanical recycling approach can reduce $CO_2$ emission, air emission with organic compounds, and waste production [5]. Mechanical recycling was carried out by the reprocessing of waste plastics with a mechanical method in order to produce new plastic products. The waste plastics were collection, cleaning, chipping, coloring or agglomeration, extrusion, and manufacturing of the end product [6]. However, the poor properties of mechanically recycled products were an obstruction to the usage of mechanical recycling [7].

In this chapter, firstly, the physical degradation of recycled plastic and its relationship with the changes in lamellar structure will be discussed. Then, with the importance of the changes in lamellar structure, the physical regeneration theory will be also introduced in order to improve the mechanical properties of mechanical recycling plastics. Then, based on the findings of physical degradation and physical regeneration theory, the revolutionary plastic mechanical recycling by using a new type of twin-screw extruder with the addition of new unit (Molten Resin Reservoir: MRR) will be introduced and showed some successful results which the mechanical properties of recycled plastics can be regenerated as similar to virgin plastics. This chapter is expected to combine the pain point of current techniques of plastic mechanical recycling process and introduce the new methods which will improve the quality of mechanical recycling process and mechanical properties of recycled plastic products.

## 2. Physical degradation of plastics, relationship with lamellar structures, and the regeneration of mechanical properties by suitable remolding conditions

Based on common sense, the properties of mechanically recycled plastics were inferior to virgin plastics in all aspects. It was not only poor physical properties such as dark color with strange odor but the mechanical properties were much degraded. In addition, the challenging works of gathering and separating to each type of plastic were one cause of the occurrence of contaminations and foreign matters. As a result, recycled plastics were brittle and became low strength and low stiffness compared to virgin plastics. From this evidence, recycled plastics had limited usability with narrow applications.

Regarding poor properties in mechanically recycled plastics, it was evidence that photodegradation, hydrolysis, and thermo-oxidation process affected the breakage of polymer chains into low molecular weight [8]. In addition, the UV-radiation and oxygen in the environment were also the most important factors for the degradation of the carbon–carbon backbone and the occurrence of chain scission in plastics [9]. As a result, the recycled plastics were become a brittle material which can be also leading to the formation of micro- and nano-plastic fragments.

In contradiction with earlier findings and a general common sense, in 2012, Yao et al. [10] investigated the molecular characteristics of virgin polypropylene (VPP) and compared with recycled polypropylene (RPP) which was obtained from byproducts of injection molding process. These byproducts were a part of VPP injection molding products namely "sprue" and "runners" (**Figure 1**) connected to

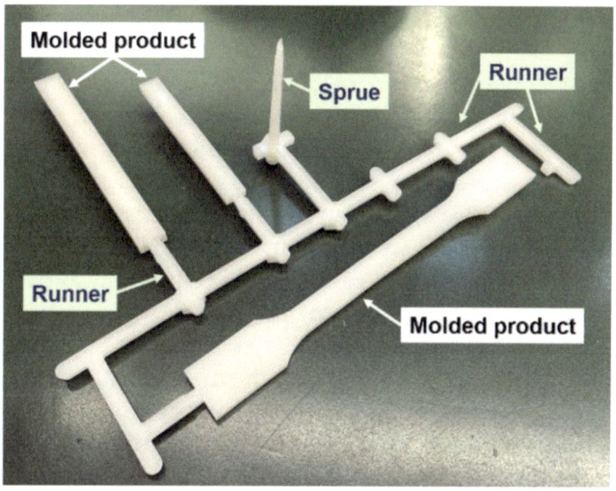


**Figure 1.**
*Part of injection molding plastic products. Sprue and Runners which were generally discarded were used as the raw material for mechanical recycling as RPP.*

the molded products which were used for tensile test and Izod impact strength test. The byproducts (sprue and runner parts) were grinding and mechanical recycling prior to remolding as RPP for characterization. It was generally found that the tensile properties of RPP were inferior to VPP due to shear deformation during mechanically recycling process. However, it is interesting to note that weight average molecular weight (Mw), number average molecular weight (Mn), and polydispersity index (PDI) of RPP were similar to VPP. This can be implied that there were no changes in chain length, chain scission, and polymer chemistry of RPP due to mechanically recycling process. In addition, the degradation of mechanical properties of RPP was strongly related to the morphological changes of inner structures. **Figure 2** showed the SEM images of the cross-section of VPP and RPP test specimen broken under liquid nitrogen. The cross-sectional area of VPP (**Figure 2(a)**) was detected as a smooth surface; however, the rough surface was detected in RPP (**Figure 2(b)**). The distinction of fracture surface morphology can be related to the difference in inner structure such as crystalline and amorphous structure, thickness, and morphology caused by the mechanical recycling process. As a result, the mechanical properties of RPP were inferior to VPP [10].

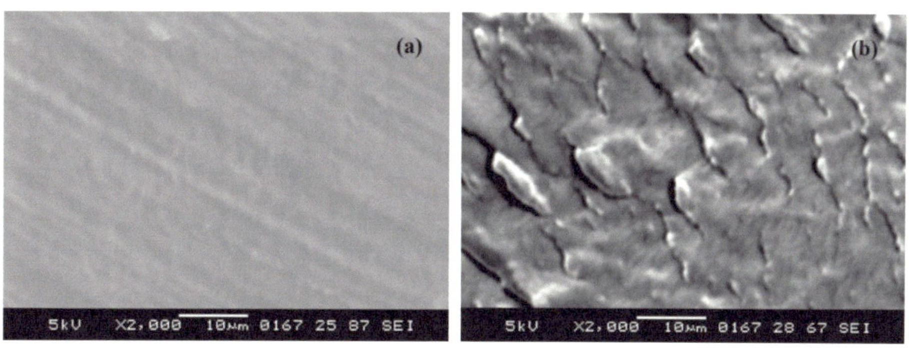


**Figure 2.**
*SEM images of the fractured surface broken under liquid nitrogen: (a) VPP; (b) RPP [10].*

In order to prove other physical properties compared between virgin and mechanical recycled plastics, Takatori et al. [11] studied the correlation of weight average molecular weight (Mw) with density, melt index, high load melt index, and mechanical properties such as yield stress and elongation at break of virgin high-density polyethylene (HDPE) and recycled HDPE originated from drinking bottle caps. It can be confirmed that main physical properties such as density, melt index, amorphous phase length, and crystal phase length of recycled HDPE were identical to virgin HDPE especially in case that they were similar in Mw [11]. From this study, it can be proved that the similarity between Mw of virgin and mechanical recycled plastics related to high possibility of comparable to other physical properties.

In order to improve the mechanical properties of recycled plastics, Tominaga et al. proposed a new promising way for improving the mechanical properties of RPP by variation of remolding conditions [12]. Their study also compared the mechanical properties between VPP and RPP which was originated from sprue and runners of injection molding products namely as "pre-RPP." The main findings from this study were shown in **Figure 3**. The general molding condition at 210°C for 2 min with slow cooling to room temperature affected low elongation at break and toughness of

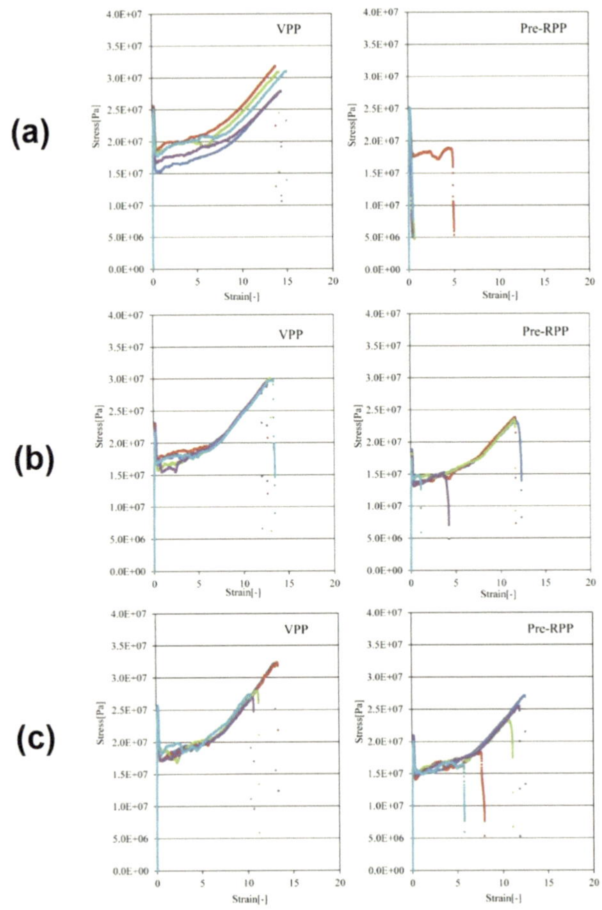


**Figure 3.**
*Stress–strain curves of VPP and pre-RPP with different molding conditions. (a) 210°C, 2 min, slow cooling; (b) 210°C, 2 min, rapid cooling; (c) 210°C, 6 min, rapid cooling [12].*

pre-RPP samples as compared to VPP (**Figure 3(a)**). As compared to the changes of cooling conditions to the rapidly cooled ice water (**Figure 3(b)**), the averaged elongation at break and toughness of pre-RPP was increased as similar to VPP. Interestingly, the elongation at break and toughness of pre-RPP can be further increased at a longer time of molding condition from 2 to 6 min as shown in **Figure 3(c)**. From these findings, it can be found that the suitable molding conditions; time and cooling method, affected the regeneration of mechanical properties especially in elongation at break and toughness of mechanical recycled RPP. In another way, this variation of molding condition has not significantly affected the changes in mechanical properties in VPP.

Tominaga et al. [13] further studied the variation of molding conditions of pre-RPP in 2015. With the variation of temperature, time, and cooling method, pre-RPP successfully improved mechanical properties, especially in tensile fracture stress, as higher than VPP. **Figure 4** showed the tensile fracture stress of VPP and pre-RPP with different molding conditions. The molding condition was varied with temperature (210, 230, and 250°C), molding time (2, 6, and 10 min), and cooling condition (slow cooling; SC, quench cooling; Q). From these results, the cooling condition by rapid cooling in ice water or quench cooling (Q) affected the increase of tensile fracture stress, especially at high molding temperature and time. Interestingly, the molding condition at 250°C for 10 min with quench cooling affected the tensile fracture stress of pre-RPP much higher than VPP with the general molding condition at 210°C for 2 min with slow cooling (SC). Based on this study, it can be also confirmed that the mechanical properties of recycled plastics were strongly related to the molding conditions. The suitable molding condition was correlated to the regeneration of mechanical properties of mechanical recycled plastics.

These findings were applied to the study by using actual recycled materials. Takenaka et al. [14] used recycled polypropylene (RPP) originating from waste containers and packaging plastics from municipal waste in Japan. The result found that the suitable molding condition especially by quench cooling affected the elongation of RPP samples. In addition, the cross-sectional of the fracture surface was totally different between the brittle sample from the remolding by slow-cooling (SC); however,

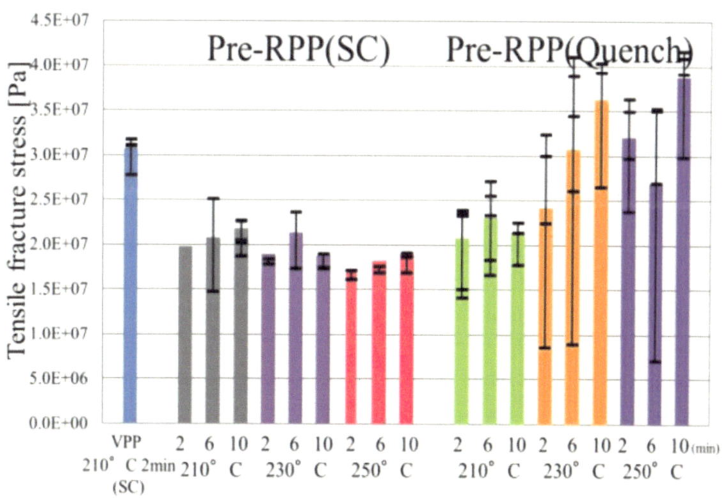


**Figure 4.**
*Tensile fracture stress of VPP and pre-RPP with different molding conditions [13].*

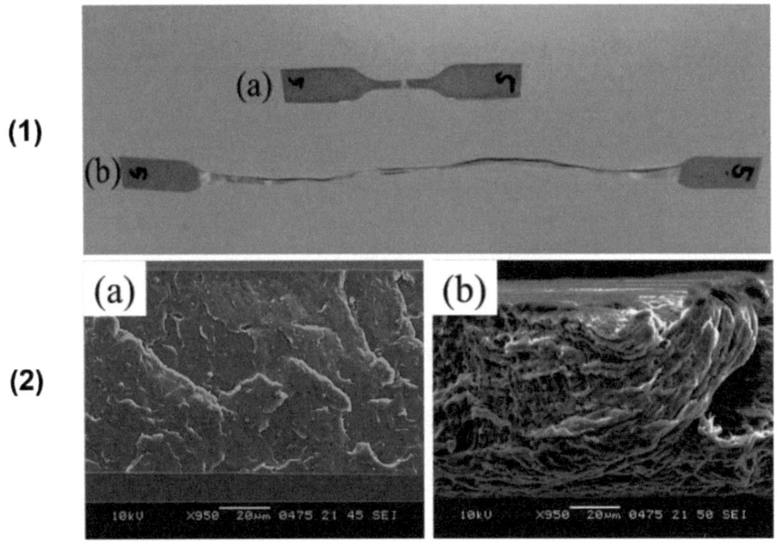


**Figure 5.**
*(1) Elongated tensile test specimen; (2) SEM images of the fractured surface broken under liquid nitrogen; (a) RPP molding at 230°C, 10 min, slow cooling; (b) 230°C, 10 min, quench cooling [14].*

the ductile and elongation surface can be detected in quenched RPP (**Figure 5**). From this study, it was found that suitable molding condition was not only related to the regeneration of mechanical properties, but it was also related to the changes of inner structure morphology in RPP originated from waste container and packaging.

For the detailed characterization of inner structure changes from suitable molding conditions in recycled plastic, a small angle X-ray scattering (SAXS) was used for characterization of thickness of long period (L), crystalline layer (Lc), amorphous layer (La), and intermediate layer (Li). **Figure 6** showed the schematic image of the lamellar structure of semicrystalline plastics such as polyethylene.

The relationship between the inner structure and mechanical properties of RPP with various molding conditions was investigated by Tominaga et al. [17]. RPP originated from byproducts of injection molding samples such as sprue and runner were molded (210 or 250°C for 2, 6, or 10 min and cooling by slow cooling or quenching) or annealed at 65°C for 2 or 8 h prior to characterization. It can be found that tensile fracture elongation showed a strong correlation with long period and thickness of amorphous layer of RPP. The shorten amorphous layer and long period with the larger number of tie molecules related to better elongation of RPP.

In the case of recycled materials from other sources; recycled plastics originated from unsorted waste containers and packaging from municipal waste in Japan, Yamasaki et al. [18] investigated the development of its mechanical properties by the optimization of molding conditions. With the complex of materials, it can be characterized that the unsorted recycled materials consisted of polyethylene (PE) and polypropylene (PP) in the ratio of 49:51. The molding conditions were compared up to 12 conditions with a variation of molding temperature (180, 210, 230, and 250°C), molding time (2, 6, and 10 min), and cooling condition (slow cooling, quench). It can be found that quench cooling affected the regeneration of toughness. Interestingly, the toughness had a strong relationship with a long period (L) as shown in **Figure 7**. Quench cooling affected the increasing of toughness with the decreasing of long period. In another way, slow cooling

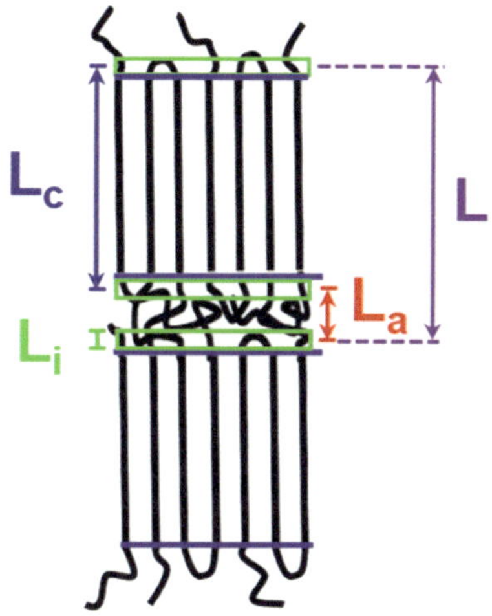


**Figure 6.**
*Lamellar structure of semicrystalline polymer; polyethylene [15, 16].*

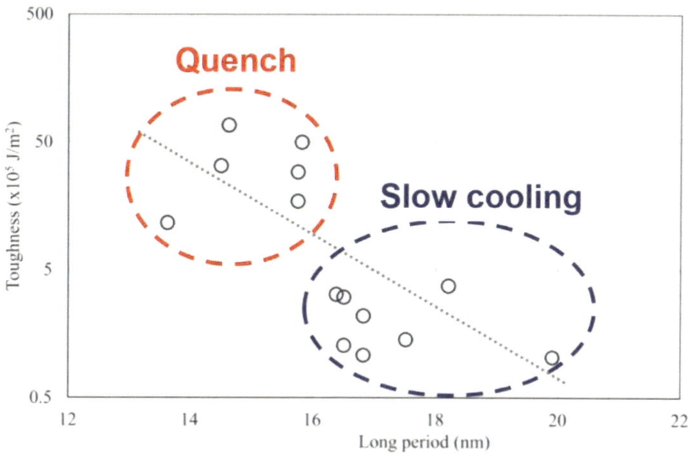


**Figure 7.**
*Relationship between long period and toughness of unsorted recycled plastics originated from municipal waste containers and packaging with different molding conditions. Remolding by quenching affected the increasing of toughness with the shortening of a long period [18].*

affected the decrease of toughness with the length of thickness over a long period. This can be also confirmed that the shortening of thickness over a long period by quench cooling affected the regeneration toughness in recycled PE and PP.

Based on the strong relationship between molding condition, mechanical properties, and lamellar structure of recycled plastics, Yao et al. [19] proposed an explanation of the physical degradation theory of semicrystalline plastics as shown in **Figure 8**.

**Figure 8(a)** shows the lamellar structure of virgin plastic which consists of crystalline as shown in blue color and tie molecules as shown in red color. There are plenty of tie molecules between crystalline lamellar. In this state, the force applied to the plastic specimen can be propagated from end to end. As a result, good tensile properties are demonstrated in virgin plastics. In contrast, as shown in **Figure 8(b)**, heating during the remolding process caused the melting of lamellar crystals. As a result, the

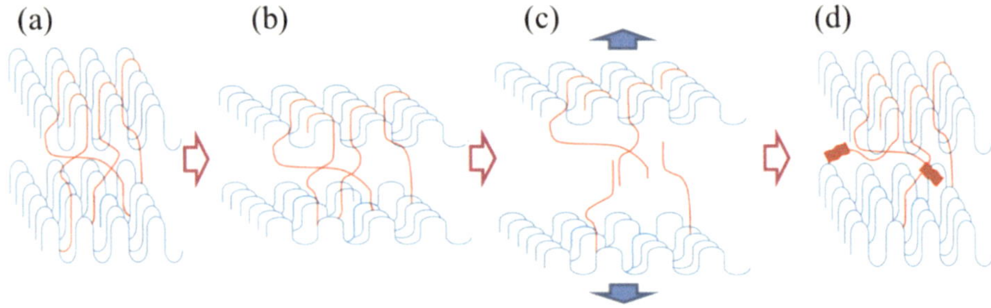


**Figure 8.**
*Schematic diagram of physical degradation mechanism. Blue color: crystalline structure; Red color: tie molecules [19].*

lamellar became thin in shape without any order. In this state, the polymer crystal had the "memory of structure" which is exhibited with a phenomenon where crystallization always occurs from the same location, unless it is when melted for a very long time [20]. Part of the retained lamellar structure shown in **Figure 8(b)** is account for this effect. The holding strength of tie molecules between lamellar crystals becomes weakened; hence, tensile deformation from the molding process results in the detachment of tie molecules from existing lamellar layers as shown in **Figure 8(c)**. However, the lamellar structures can be reconstructed from this state by cooling, after molding. The detached tie molecules become dangling chains without returning to the original lamellar structure. Therefore, molded products consisted of a few tie molecules, as shown in **Figure 8(d)**. This structure cannot sufficiently propagate force from one end of a specimen to the other, resulting in the severe degradation of mechanical properties. In short, physical degradation is caused by the transformation of lamellar structure with inferior mechanical properties from heat and shear accompanying the remolding process condition [19].

Based on the study on the relationship between suitable molding conditions and the regeneration of mechanical properties of semi-crystalline plastic such as RPP, then the studies were extended to amorphous plastics; polystyrene (PS). Oda et al. [21, 22] studied the development of mechanical properties of virgin and recycled PS by variation of molding temperature and holding time. As a result, molding conditions at 230°C, 40 MPa for 2 min affected the improvement of mechanical properties in PS. Interestingly, the new peaks can be detected in the wide-angle X-ray scattering (WAXS) profile which can be implied by the changes in the inner structures of the amorphous polymer due to the distinction of molding condition. In addition, it can be also concluded that the suitable molding condition for regeneration of mechanical properties was not only able to apply to semi-crystalline plastic but it can be also applied to the amorphous polymer such as PS.

## 3. Physical degradation of plastics from shear deformation: the current obstacle of plastic mechanical recycling

Extrusion process which was the main approach in plastic mechanical recycling consisted of shear deformation from screw in mixing zone and conveying zone. As a result, mechanical properties of recycled products were degraded which was caused by the memory of shear history which remained in the structure of plastics. Many studies have focused on the shear deformation and crystalline structure of plastics.

Liu et al. [23] evaluated shear crystallization and changes in the crystalline structure of HDPE using synchrotron X-ray techniques. It can be found that shear accelerated the speed of crystallization and the formation of crystalline orientation along the flow direction. Abad et al. [24] evaluated the tensile properties of HDPE and low-density polyethylene (LDPE) after several extrusion cycles. The tensile properties were degraded with the greater number of extrusion cycles. Interestingly, the addition of antioxidation was able to reduce the effect of degradation.

Kaneyasu et al. [25] investigated the mechanism of steady shear deformation of HDPE. Mechanical recycling process was performed by steady shear treatment by using a cone and plate rheometer. The steady shear treatment was varied from 0 to 100 $s^{-1}$ at 180 °C for 10 min. Then, the product was remolded as a thin film with a thickness of 100 μm for further characterization. As a result, steady shear treatment at 100 $s^{-1}$ affected the degradation of elongation at break and Young's modulus from those of virgin HDPE around 27.9% and 27.8%, respectively. Interestingly, it can be found that the degradation of elongation at break from steady shear had a strong correlation with the decreasing the thickness of the long period, the core crystalline layer, and the intermediate layer. In addition, the surface morphology of HDPE treated by 100 $s^{-1}$ showed the creation of crystalline orientation which is also related to the decreasing of elongation at break in HDPE. On the other hand, it can be confirmed that the degradation of elongation at the break of HDPE is caused by the physical degradation from the changes of lamellar structure and morphology. It was not caused from the changes of polymer chemistry due to there were no changes in molecular weight and molecular weight distribution of shear-treated HDPE as compared to its virgin HDPE. From this study, it can be evidence that steady shear treatment which was generally detected in the mechanical recycling process affected the degradation of mechanical properties which can be also related to the changes of thickness of lamellar structure and mesoscale surface morphology of HDPE (**Figure 9**).

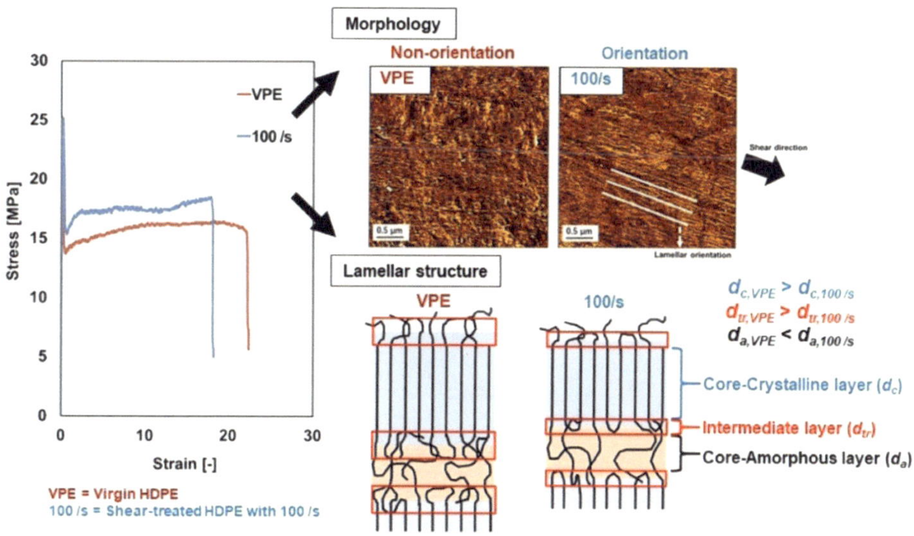


**Figure 9.**
*The comparison of tensile properties, morphology, and thickness of lamellar structure of virgin HDPE (VPE) and steady shear-treated HDPE at 100 $s^{-1}$. The elongation at break of steady shear-treated HDPE was decreased from its VPE with the creation of crystalline orientation at shear direction and the reduction of thickness of core crystalline layer and intermediate layer [25].*

## 4. Physical regeneration theory and the revolutionary plastic mechanical recycling process

From the knowledge of physical degradation theory, it can be further thought about the development of mechanical properties of plastics by the relationship with the regeneration of lamellar structure. Yao et al. [26] have been also proposed the physical regeneration theory of semicrystalline plastics as shown in **Figure 10**.

This physical degradation and regeneration mechanism can be theoretically explained by using **Figure 10**, which shows the energy levels and states of solid and molten polymers. First, in the crystalline state, the crystalline lamellar structure consisting of polymer chains is incomplete and there are many tie molecules, as shown in **Figure 10(a)** with a low energy level. Therefore, the polymer chains are constantly moving toward rearrangement in this state. The temperature at which the crystallization rate is the fastest in crystalline polymers is considered to be the temperature near the middle of glass transition temperature which molecular motion completely stops and the crystalline melting point. Therefore, the temperature at crystallization rate is the fastest is approximately 70°C for PP and approximately 0°C for PE. In other words, in many plastic products, which are recognized to be in a solid state when used, the inner polymer chains are in a metastable/non-equilibrium state, in which the inner polymer chains are constantly moving toward a stable state of energy. If the thermal history is sufficient for progression of crystallization, it is considered that the state in **Figure 10(b)** is infinitely approached. In this state, since the number of tie molecules is small, the transmission of force in the system is interrupted and the elongation property is greatly deteriorated. The proponents actually confirmed that the specimens in which the crystallization was accelerated by cooling and solidifying virgin PP very slowly became markedly brittle.

On the other hand, in the molten state, the energy level is lower in the compatible state where polymer chains are entangled with each other shown in **Figure 10(d)** than in the phase-separated state shown in **Figure 10(c)**. However, when the lamellar structure in the state shown in **Figure 10(d)** is subjected to a strong shear deformation

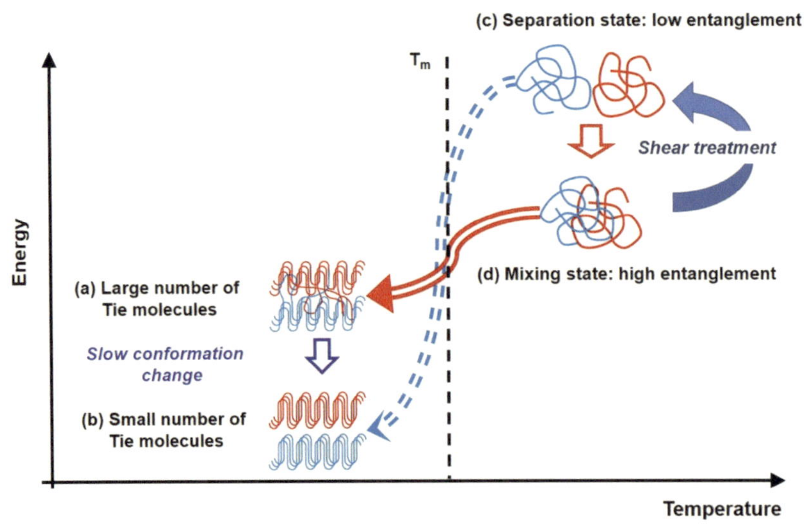


**Figure 10.**
*Schematic diagram of physical degradation and regeneration mechanism of semicrystalline plastics [26].*

due to the extrusion and molding process, the entanglement between polymer chains was decreased and the system assumes the state shown in **Figure 10(c)**. When the state shown in **Figure 10(c)** is cooled and solidified, the entanglement between the polymer chains has already decreased, so the state shown in **Figure 10(b)** is lower than the state shown in **Figure 10(a)**, which has excellent mechanical properties. This is the reason why the physical properties of recycled plastics with a history of molding have deteriorated and shown in **Figure 10(b)**. In addition, the system in which the molecules are separated from each other and melted as shown in **Figure 10(c)**. However, it is possible to return to a more stable energy state shown in **Figure 10(d)** by leaving it in a molten state for a long time. In other words, physical properties of plastics can be "physically regenerated" by molding through a process that utilizes the self-regenerating ability of this polymer. However, once a stable state is established in a crystalline polymer, there is a phenomenon known as a "memory effect", in which the polymer chain retains its structure for a long time [27]. Therefore, even if entanglement is formed by placing it in a molten state, it can be easily disentangled during the cooling process if the temperature degree is low. Conversely, if such measures are taken, there is a possibility that the physical properties can be stably reproduced.

Based on this physical regeneration theory, the creation of a new type of extrusion process was exhibited. Phanthong et al. [28] studied the development of mechanical properties and crystalline conformation of recycled polypropylene by re-extrusion using a new type twin-screw extruder. This was the establishment of a new unit which was a blank space added into the end of a twin-screw extruder, namely a "Molten Resin Reservoir: MRR". This study compared the mechanical properties and inner structures of virgin polypropylene (VPP) and recycled polypropylene (RPP) extruded by using a twin-screw extruder with an additional MRR as shown in **Figure 11**. The tensile properties, crystalline type, inner structure conformation, and molecular weight distribution were also characterized.

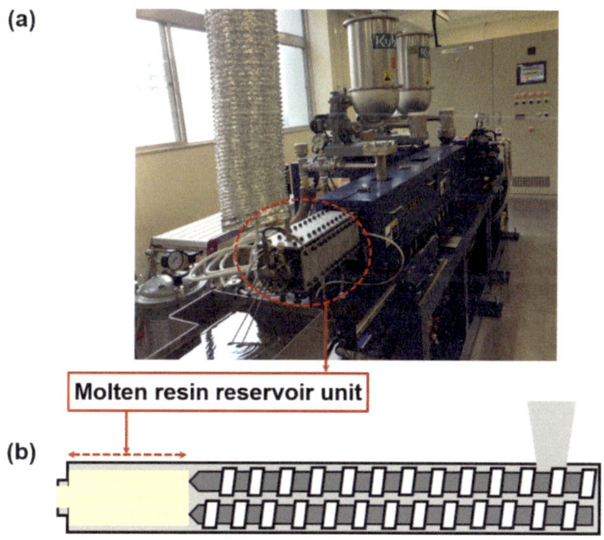


**Figure 11.**
*(a) New type of twin-screw extruder with additional MRR unit; (b) schematic image of MRR unit which was a blank space connected at the end of screw mixing zone of general extruder [28].*

As a result, the tensile properties of RPP-extrusion were improved and were similar to those of VPP (**Figure 12**). In addition, the crystalline conformation of RPP-extrusion was similar to that of VPP by increasing the ratio between the helix and parallel band which could be attributed to the improvement of tensile properties. In addition, the molecular weight distribution of RPP-extrusion was similar to its original sample which can be implied to the fact that the re-extrusion process by the twin-screw extruder with the addition of MRR did not affect changes in the chain length and chain structure of RPP. This study succeeded in regenerating the tensile properties and inner structures of recycled PP.

The use of new type extrusion with MRR was also studied with the unsorted recycled plastics originating from waste container and packaging plastics in municipal waste in Japan. Yamasaki et al. [18] studied the re-extrusion process with various conditions of the existence of MRR, extrusion temperature (200, 230, and 250°C), screw speed (100 and 200 rpm), and cooling method (room temperature and ice water). As a result, the use of MRR affected the increase of elongation at the break of samples. In addition, the optimized condition for the unsorted recycled plastics derived from waste container and packaging was re-extrusion with the addition of MRR at 200°C, 100 rpm, and cooling in ice water (**Figure 13**).

The addition of MRR not only succeeded in the regeneration of mechanical properties in RPP, recycled polyethylene (RPE) was also tested in the re-extrusion process with MRR. Okubo et al. [29] studied the effect of new type twin-screw extruder with the addition of MRR for the regeneration of mechanical properties of RPE derived from plastic containers and packaging. As a result, the elongation at break of RPE re-extrusion by MRR was seven times higher than that of its original RPE (**Figure 14**).

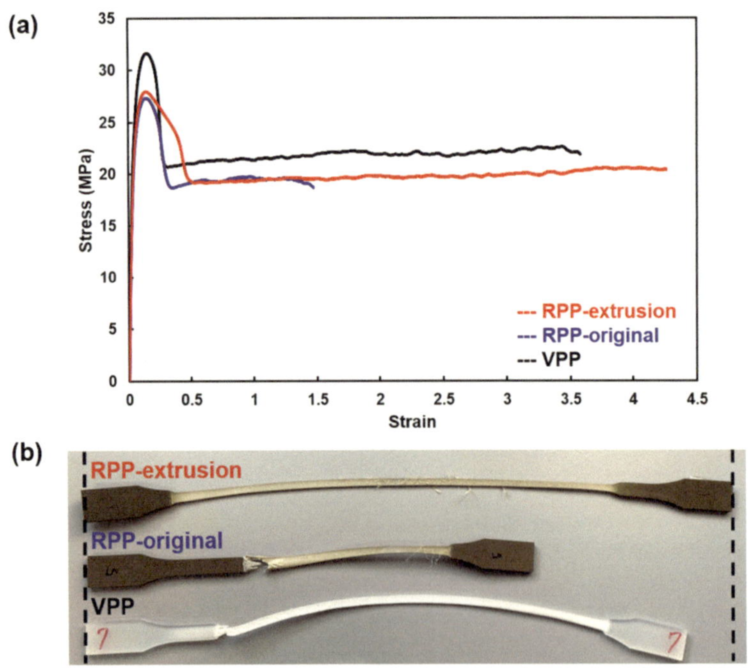


**Figure 12.**
*(a) Stress–strain curve of VPP, RPP-original sample without re-extrusion, and RPP-extrusion by using a new type twin-screw extruder with additional MRR unit; (b) photo images of the elongated tensile test specimen. RPP-extrusion had been more elongated than its original sample (RPP-original) and VPP [28].*

| Sample No. without reservoir | Sample No. with reservoir | Pelletize temperature (℃) | Rotation speed of cylinder (rpm) | Water bath temperature (℃) |
|---|---|---|---|---|
| ① | with-D | 200 | 100 | Room temp. |
| ② | with-D | 200 | 200 | Room temp. |
| ③ | with-D | 230 | 100 | Room temp. |
| ④ | with-D | 230 | 200 | Room temp. |
| ⑤ | with-D | 250 | 100 | Room temp. |
| ⑥ | with-D | 250 | 200 | Room temp. |
| ⑦ | with-D | 200 | 100 | Ice water |
| ⑧ | with-D | 200 | 200 | Ice water |
| ⑨ | with-D | 230 | 100 | Ice water |
| ⑩ | with-D | 230 | 200 | Ice water |
| ⑪ | with-D | 250 | 100 | Ice water |
| ⑫ | with-D | 250 | 200 | Ice water |

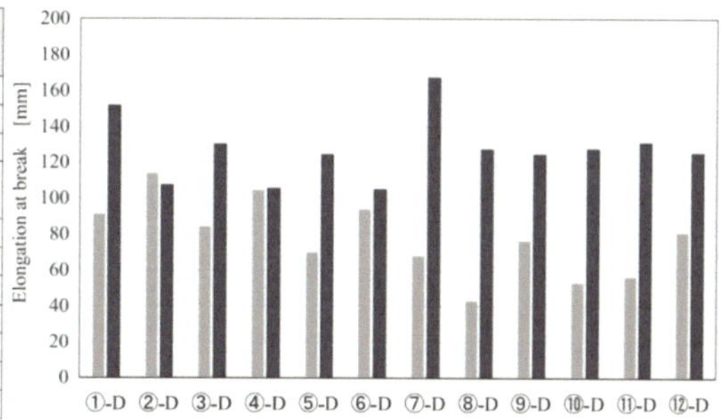


**Figure 13.**
*Elongation at break of recycled plastics derived from unsorted waste container and packaging and re-extrusion with various extrusion conditions [18].*

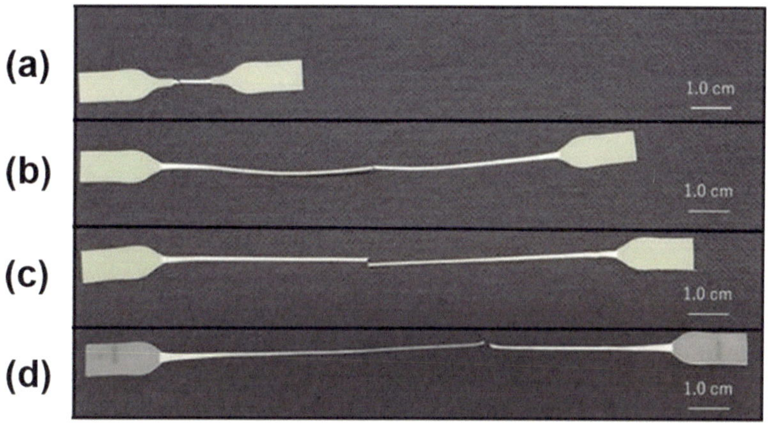


**Figure 14.**
*Photo images of the elongated tensile test specimen: (a) original RPE; (b) RPE re-extrusion with general extruder; (c) RPE re-extrusion by MRR; (d) VPE. RPE re-extrusion by using MRR was elongated higher than its original RPE around seven times and also similar to VPE [29].*

For the consideration of the changes in lamellar structure from the re-extrusion by MRR, the lamellar shape of RPE was significantly changed from a distorted stripe-like lamellar structure to a stripe-like lamellar structure which was similar to VPE as shown in **Figure 15**.

Based on the changes in lamellar morphology, the effects of MRR on microstructure of RPE was shown in **Figure 16**. It can be found that all samples exhibited different

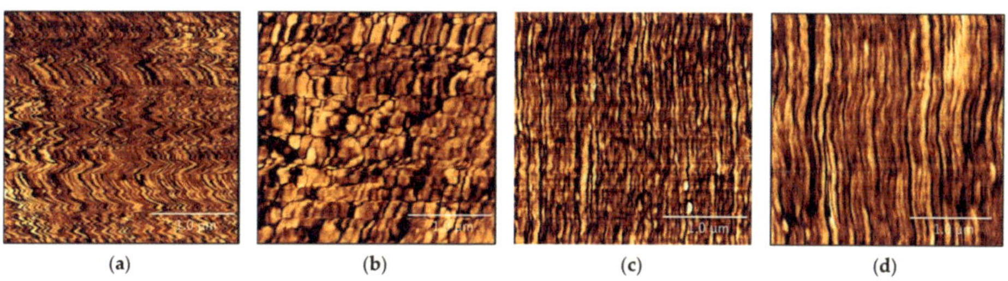


**Figure 15.**
*AFM phase images: (a) original RPE; (b) RPE re-extrusion with general extruder; (c) RPE re-extrusion by MRR; (d) VPE [29].*

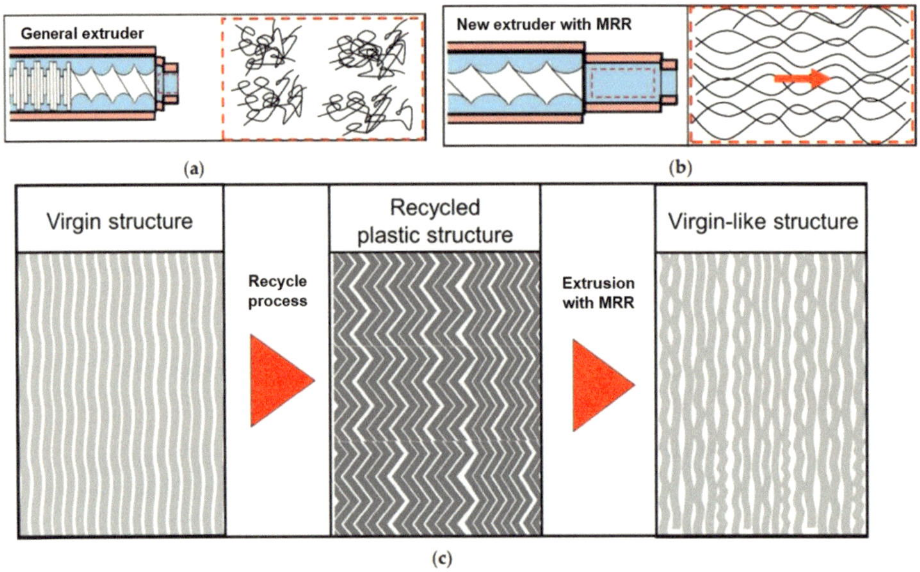


**Figure 16.**
*Schematic diagram of the changes of polymer chain in the molten state: (a) general extruder; (b) new type extruder with the addition of MRR unit; (c) schematic images of the changes of lamellar morphology of virgin plastics by general recycling process. The lamellar structure was regenerated to be a virgin-like structure by re-extrusion using the addition of MRR unit [29].*

lamellar structures depending on the processing method. This indicated that the structure of the molten polymer is reflected in the structure of the molding as a result of melt-memory effects during polymer crystallization [30]. For the re-extrusion in general, extruder without MRR, the molten polymer was mixed in the kneading zone. As a result, the molten polymer had a dispersed structure as shown in **Figure 16(a)**. In contrast, in the re-extrusion by using MRR, the flow of the molten polymer was laminar in MRR resulting in the alignment of the molten polymer chains in the flow direction as shown in **Figure 16(b)**. As a result, re-extrusion of RPE by MRR can improve the distorted lamellar structure of RPE pellet yielding a virgin-like structure (**Figure 16(c)**). Consequently, the mechanical properties were also regenerated as virgin-like properties.

## 5. Conclusion and future prospects

The degradation of mechanical properties of plastics was revealed that was caused by physical degradation. From the plastic processing by heat and shear from general extrusion or injection molding process affected the changes in lamellar structure, morphology, size, the amount of entanglement, and tie molecules. In another way, physical regeneration can be established by the suitable mechanical processing condition and the relaxation of memory from the molten state of polymer chains. The establishment of molten resin reservoir unit (MRR) which was a blank space connected at the end of the screw unit of the general extruder affected the relaxation and removable of the shear history of molten polymer. As a result, the lamellar structure and morphology can be regenerated as similar to its virgin materials which also affected to the regeneration of the elongation at break. These findings have already succeeded for recycle polyethylene and recycled polypropylene derived from plastic waste containers. From this study, it can be concluded that the mechanical recycling of plastics can be revolution which can be increased the quality of recycled plastic products which can be prolonged the lifetime used and decreased the amount of single used plastic waste in environment.

Future prospects of this work will be focused on the extension of plastic types and composite materials. In addition, the modification of extrusion condition and MRR will be also further studied in order to boost up the quality of the mechanically recycled plastics.

## Acknowledgements

This study is based on results obtained from a project no. JPNP20012 commissioned by the New Energy and Industrial Technology Development Organization (NEDO) of Japan. This research was supported by the Environment Research and Technology Development Fund (3-1705) of Environmental Restoration and Conservation Agency of Japan.

## Conflict of interest

The authors declare no conflict of interest.

## Author details

Patchiya Phanthong[1*] and Shigeru Yao[1,2]

1 Research Institute for the Creation of Functional and Structural Materials, Fukuoka University, Fukuoka, Japan

2 Faculty of Engineering, Department of Chemical Engineering, Fukuoka University, Fukuoka, Japan

*Address all correspondence to: patchiya@fukuoka-u.ac.jp

## References

[1] Pakhomova S, Zhdanov I, van Bavel B. Polymer type identification of marine plastic litter using a miniature near infrared spectrometer (microNIR). Applied Sciences. 2020;**10**:8707. DOI: 10.3390/app10238707

[2] Chamas A, Moon H, Zheng J, Qiu Y, Tabassum T, Jang J, et al. Degradation rates of plastics in the environment. ACS Sustainable Chemistry & Engineering. 2020;**8**:3494-3511. DOI: 10.1021/acssuschemeng.9b06635

[3] Plastic Waste Management Institute. An introduction to plastic recycling. 2019. Available from: http://www.pwmi.or.jp/ei/plastic_recycling_2019.pdf [Accessed: 31 December 2020]

[4] Alassali A, Picuno C, Bébien T, Fiore S, Kuchta K. Validation of near infrared spectroscopy as an age-prediction method for plastics. Resources, Conservation and Recycling. 2020;**154**:104555. DOI: 10.1016/j.resconrec.2019.104555

[5] Perugini F, Mastellone M, Arena U. A life cycle assessment of mechanical and feedstock recycling options for management of plastic packaging wastes. Environmental Progress. 2005;**24**:137-154. DOI: 10.1002/ep.10078

[6] Francis R, editor. Recycling of Polymers: Methods, Characterization and Applications. Weinheim, Germany: Wiley-VCH; 2016. p. 288

[7] Ragaert K, Delva L, Geem K. Mechanical and chemical recycling of solid plastic waste. Waste Management. 2017;**69**:24-58. DOI: 10.1016/j.wasman.2017.07.044

[8] Samir Ali S, Elsamahy T, Koutra E, Kornaros M, El-Sheekh M, Abdelkarim EA, et al. Degradation of conventional plastic wastes in the environment: A review on current status of knowledge and future perspectives of disposal. Science of The Total Environment. 2021;**771**:144719. DOI: 10.1016/j.scitotenv.2020.144719

[9] Gewert B, Plassmann MM, MacLeod M. Pathways for degradation of plastic polymers floating in the marine environment. Environmental Science: Processes & Impacts. 2015;**17**:1513-1521. DOI: 10.1039/C5EM00207A

[10] Yao S, Tominaga A, Fujikawa Y, Sekiguchi H, Takatori E. Inner structure and mechanical properties of recycled polypropylene. Nihon Reoroji Gakkaishi. 2013;**41**(3):173-178. DOI: 10.1678/rheology.41.173

[11] Takatori E, Shimura T, Yao S, Shindoh Y. Dependencies of material properties on averaged molecular weight of a recycled high-density polyethylene. Nihon Reoroji Gakkaishi. 2014;**42**(1):39-43. DOI: 10.1678/rheology.42.39

[12] Tominaga A, Sekiguchi H, Nakano R, Yao S, Takatori E. Advanced recycling technology of pre-consumer waste polypropylene. Kobunshi Ronbunshu. 2013;**70**:712-721. DOI: 10.1295/koron.70.712

[13] Tominaga A, Sekiguchi H, Nakano R, Yao S, Takatori E. Thermal process-dependence of the mechanical properties and inner structures of pre-consumer recycled polypropylene. AIP Conf. Proc. 2015;**1664**:150011. DOI: 10.1063/1.4918507

[14] Takenaka N, Tominaga A, Sekiguchi H, Nakano R, Takatori E, Yao S. Creation of advanced recycle

process to waste container and packaging plastic – Polypropylene sorted recycle plastic case. Nihon Reoroji Gakkaishi. 2017;**45**(3):139-143. DOI: 10.1678/rheology.45.139

[15] Lewis TJ, Llewellyn JP. Electrical conduction in polyethylene: The role of positive charge and the formation of positive packets. Journal of Applied Physics. 2013;**113**(223705). DOI: 10.1063/1.4810857

[16] Hsu Y, Truss R, Laycock B, Weir M, Nicholson T, Garvey C, et al. The effect of comonomer concentration and distribution on the photooxidative degradation of linear low density polyethylene films. Polymer. 2017;**119**:66-75. DOI: 10.1016/j.polymer.2017.05.020

[17] Tominaga A, Sekiguchi H, Nakano R, Yao S, Takatori E. Relationship between the long period and the mechanical properties of recycled polypropylene. Nihon Reoroji Gakkaishi. 2017;**45**(5):287-290. DOI: 10.1678/rheology.45.287

[18] Yamasaki N, Takenaka N, Tominaga A, Michiue T, Takatori E, Sekiguchi H, et al. Molding condition dependence of mechanical properties of unsorted recycled waste container and packaging plastics. Journal of the Japan Society of Material Cycles and Waste Management. 2019;**30**:122-131. DOI: 10.3985/jjsmcwm.30.122

[19] Yao S, Yamasaki N, Phanthong P, Yamashita K, Ueno Y, Michiue T, Takatori E. Novel material recycle process based on the physical degradation and physical regeneration theory. In: Proceedings of the International Conference on Advanced and Applied Petroleum, Petrochemicals, and Polymers (ICAPPP2018); 18-20 December 2018; Bangkok, Thailand; 2018. pp. 111-118

[20] Ichikawa Y, Iizuka Y, Katto T. Crystallization kinetics of polyphenylene sulfide: Effect of molecular weight. Kobunshi Ronbunshu. 1990;**47**(1):73-77. DOI: 10.1295/koron.47.73

[21] Oda N, Sekiguchi H, Nakano R, Yao S. Molding history dependence of the mechanical properties of recycled amorphous plastics. Nihon Reoroji Gakkaishi. 2016;**44**(1):47-53. DOI: 10.1678/rheology.44.47

[22] Oda N, Tominaga A, Sekiguchi H, Nakano R, Yao S. Development of an internal structure by amorphous polymer in the melt state. Nihon Reoroji Gakkaishi. 2017;**45**(2):101-105. DOI: 10.1678/rheology.45.101

[23] Liu Y, Gao S, Hsiao B, Norman A, Tsou A, Throckmorton J, et al. Shear induced crystallization of bimodal and unimodal high density polyethylene. Polymer. 2018;**153**:223-231. DOI: 10.1016/j.polymer.2018.08.020

[24] Abad MJ, Ares A, Barral L, Cano J, Díez FJ, García-Garabal S, et al. Effects of a mixture of stabilizers on the structure and mechanical properties of polyethylene during reprocessing. Journal of Applied Polymer Science. 2004;**92**(6):3910-3916. DOI: 10.1002/app.20420

[25] Kaneyasu H, Phanthong P, Okubo H, Yao S. Investigation of degradation mechanism from shear deformation and the relationship with mechanical properties, lamellar size, and morphology of high-density polyethylene. Applied Sciences. 2021;**11**:8436. DOI: 10.3390/app11188436

[26] Yao S, Phanthong P. Innovative material recycling process utilizing the self-resilience ability of plastics. Kagaku Kogaku. 2021;**85**(3):157-159

[27] Hoffman J, Davis G, Lauritzen J. The rate of crystallization of linear polymers with chain folding. In: Hannay NB, editor. Treatise on Solid State Chemistry. 1st ed. Boston: Springer; 1976. pp. 497-614. DOI: 10.1007/978-1-4684-2664-9_7

[28] Phanthong P, Miyoshi Y, Yao S. Development of tensile properties and crystalline conformation of recycled polypropylene by re-extrusion using a twin-screw extruder with an additional molten resin reservoir unit. Applied Sciences. 2021;**11**:1707. DOI: 10.3390/app11041707

[29] Okubo H, Kaneyasu H, Kimura T, Phanthong P, Yao S. Effects of a twin-screw extruder equipped with a molten resin reservoir on the mechanical properties and microstructure of recycled waste plastic polyethylene pellet moldings. Polymers. 2021;**13**:1058. DOI: 10.3390/polym13071058

[30] Muthukumar M. Communication: Theory of melt-memory in polymer crystallization. The Journal of Chemical Physics. 2016;**145**:031105. DOI: 10.1063/1.4959583

**Chapter 5**

# Recycling Gap, Africa's Perspective for Sustainable Waste Management

*Florence Akinyi Ogutu and Bessy Kathambi*

**Abstract**

Africa is a rising continent with a lot of development taking place having a youthful robust growing population. Africa's urbanization is projected to double in the next decade with an equally increasing population in urban and cities surroundings. A growing population also intimates that waste generation will also double if not triple from the developments and human activities associated with a growing urban populace. Waste generation from a global level is expected to double by 2050 with cities and urban centers being the highest contributors to waste generation. Thus, Africa will require measures and policies that will address the future of sustainable waste management in its entirety. In this regard, Africa needs to take stock of their waste management infrastructure and highlight the gaps in existence. One of the problems that are crosscutting in Africa is the gap in full realization of the potential of recycling of waste and the economic and environmental gains attributed to recycling. In Africa, less than 10% of the countries having recycling plants that operate optimally as well as have infrastructure that can sustain proper waste management from financial to personnel. This chapter outlines the gap in recycling.

**Keywords:** waste management, recycling, gap, Africa continent perspective, sustainable waste management

## 1. Introduction

The potential of Africa as a continent explicates the need for a robust waste management infrastructure if they are to maintain a clean environment overall [1, 2]. Africa being the second-largest continent characterized with a growing youthful population and increased economic development, waste generation will be inevitable as a result as of rapid urbanization and population increase [3]. In Africa, waste management could pose serious threats to humanity if not addressed sustainably due to change in consumption habits leading to high consumption of resources resulting in an increased waste generation [1, 4]. The rapid urbanization of Africa explicates the need for urgency in reviewing and equipping the waste management infrastructure by adopting the 3Rs (Reduce, Re-use, and Recycle) paying more attention on recycling [5, 6].

The tenet of Recycling comprises of collection and processing of waste into new products with the focus of having zero waste [2, 7]. Recycling equips people with an environmental ethic of avoiding the littering culture and sensitization of how a clean environment should be [7, 8]. The benefit of recycling range from the reduction of

 IntechOpen

pollution and global warming, wealth creation, and ultimately lessens the amount of waste taken to landfills [3, 9].

Africa's emergence in recycling is characterized by poverty, unemployment, and socioeconomic needs that arise from the demands of public and private demands especially for those in urban centers and cities [10, 11]. Interestingly, the waste generated in most urban centers and cities in Africa is recyclable waste however only a very small percentage have adopted recycling [12, 13]. Further, recycling is conducted by the informal waste pickers to subsidize their livelihoods from actively recovering valuable resources in the waste to sell to private sector [14].

Consequently, the use of informal sector to conduct recycling renders waste generated from urban centers and cities unproductive since they lack the capacity and technology to fully recycle [2, 5]. In this regard, it is estimated that approximately half of the waste material generated in Africa remains uncollected within Africa's cities and towns, where it remains dumped on sidewalks, open fields, storm water drains, and rivers leading to mushrooming dumpsites [15].

Worth noting is that the primary causes of inadequate waste disposal and management in Africa are envisaged in weak strategic, institutional, and organizational structures which are perpetrated by limited skills that are essential to waste material management; inadequate budgets; feeble legislation; and lack of enforcement necessary for waste management [3, 7]. Additionally, low public awareness, increasing corruption, and conflict lead to political instability in various African countries; and generally a lack of political will to deal with waste material disposal and management more so recycling [6, 16].

African countries conventionally have managed waste disposal in landfills, indiscriminate dumping, open burning, and recycling where they deem fit [3, 15]. Unfortunately, the health and environmental impacts projected by increased levels of waste in Africa, exposure of the public to waste sites, are detrimental and raise concerns across the African continent [3, 16]. Environmental impacts include the fact that the dumped waste material may find its way into water bodies through leaching over time, sometimes into groundwater bodies, and thus causing water pollution further affecting biodiversity [2, 17].

However, there is a window of hope, spearheaded by the African Union (AU), in their strategic socio-economic transformation framework for the continent, titled; Africa We Want (Agenda 2063), 2015, one of the targets is achieving a recycling rate of 50% of urban waste by 2023 and growing urban waste recycling industries. South Africa is a model state of recycling in Africa, at a rate of 30% on solid waste such as plastic, and bottles [5, 6].

Developed countries have made tremendous progress in recycling which has helped them manage the problem of waste management [1, 18]. This is done through controlling the waste generation, use of technology and research in energy recovery, recycling, reducing waste taken to landfills, promoting of 3R concept (Reduce, reuse, and recycle), and encouraging the use of recycled products by the public which impacting negatively on the environment. Germany, Sweden, and Norway are model countries in recycling; their recycling rate is at 70–85% [19]. Thus, the gap in recycling in Africa is multifaceted and requires collaborative engagement and partnership to enhance recycling [7, 20].

### 1.1 Recycling and sustainable development in Africa

Sustainable development according to the Brundtland Commission encompasses environment, social, and economic intersections for there to be sustainability and more so in waste management [11]. The role of recycling in sustainable solid waste management provides opportunities to enhance sustainable development envisioned by the commission [21, 22]. The issue of solid waste management in Africa is portrayed as a major developmental challenge whose serious consequences impact severely environmental quality, public health, fisheries, agriculture, and sustainable development [16, 18]. The introduction of Sustainable Development Goals (SDGs) gives African countries opportunities to enhance recycling as a key component of sustainability through improvement of solid waste infrastructures, skills, and expertise that are necessary to tackle the amount and complexity of solid waste being produced [23, 24]. The potential that recycling has in accelerating sustainable development can be a reality in that of the world's biggest dumpsites, Africa has 19 of them out of 50 [16, 18]. Further, the high population growth in Africa and rapid urbanization projected to double by 2050 and waste generation exceeding 160 million tons by 2025 explicates why recycling can be the stepping stone to sustainable development [9, 25].

Linking recycling with SDGs, Africa's poor waste management practices in particular the widespread dumping of wastes in water bodies and uncontrolled dump sites, aggravates the problems of generally low sanitation levels across the continent [8, 26]. Thus, water bodies can get reprieve if recycling was upscaled and better environmental ethics be disseminated to the communities [11, 27].

Changing the perspective of the rapid urbanization in Africa, recycling brings in a new avenue of job creation and improvement of the economic outlook for both rural and urban populations [25, 28]. Additionally, as urbanization increases, the need for different infrastructure development arises thereby giving rise for the inclusion for sustainable solid waste management practices which could foster the potential of recycling [6, 7].

The focus on having sustainable global cities which are anchored on SDG 11 ensures that the potency of recycling can be adopted in land use planning which will also cater for the growing percentage of informal settlements around these cities [7, 29]. Worth noting is that the growth of informal settlements provides workforce needed in recycling plants where in the long run it improves their sources of livelihoods underpinning SDG1 on ending poverty [30, 31].

Interestingly, waste management infrastructure is largely nonexistent in rural areas of Africa which gives opportunities for better avenues of managing their wastes as they get to be urbanized [10, 23]. Improvements in infrastructure will aid in combating the high costs of health services, and hence alleviate poverty, and reduce rural-urban migration from a clean environment in line with SDG 3, 15, and 1 [30, 32].

Partnerships in reducing the gap between waste management policy and legislation and actual waste management practices can be enhanced through participatory, inclusive stakeholder engagements that foster the target points of SDG17 [10, 25]. Partnership and networking will assist in sharing technological know-how, best practices that are sustainable and best fits within varied contexts [11, 20].

In Africa, the adoption of technology is fast growing more so in the use of ICT, which by itself gives further rise to E-waste, an addition to the other waste. Adopting recycling practices especially in E-waste provides for employment opportunities and manages the consumption trends in a sustainable manner [7, 23]. The understanding of the changing lifestyles and consumption patterns of the growing urban middle class highlights increasing the complexity and composition of waste streams in Africa necessitating the inclusion of recycling for sustainable development [18, 21].

Sustainable development looks at a powered economy and therefore energy is an important aspect for growth of urban and peri-urban areas for which recycling becomes useful in generation of green energy such as Biogas which is environmentally friendly [5, 32]. Recycling enhances Biogas and compost production from organic waste fractions as a best practice, and progress is being made by developing countries attaining SDG 7 on energy [30, 31].

Moreover, these innovations in waste management for Africa and globally through recycling are meeting many of the SDGs by minimizing greenhouse gas emissions which are helping to alleviate the climate crisis, conserving oceans and land to preserve marine and wildlife, and ensuring healthy lives and promoting well-being. These benefits from the recycling of waste enhance the capacity of countries in attainment of SDG 13, 14, 15, 3, 4, 2, 1, and 17 specifically and contributing extensively on the other SDGs [5, 21, 33]. Additional benefits from innovations in plastic waste management one of the good examples of what recycling can do include energy conservation, reduced petroleum use, and reduced landfill use [7, 33].

Interlinkages of recycling and sustainable development give hope of a better future for generations to come and answers the many questions arising on humanity existence and its role in restoration of earth [33]. Recycling has power to transform Africa to a green continent which has exudes in environmental sustainability in waste management.

## 2. Recycling in Africa, Case of Rwanda and South Africa

### 2.1 Recycling in Rwanda

Environmental protection is the pillar of waste management system in Rwanda and it is a collective responsibility involving all the stakeholders; private sector, the public, and Government agencies. Policies on solid waste management are intertwined with aspects of Rwandan culture and traditional practices. Umuganda (community work), cleaning the environment, has taught environmental values and ethics, thus creating a sustainable system [22–25].

Globally recycling is used to address the challenges of waste management especially controlling the amount of waste taken to landfills and reducing GHG emissions from uncontrolled dumping and open burning, which impact negatively on the environment and human health [3, 15].

Many countries increasingly employ sustainable waste removal and management technologies in Africa by exploiting evolving waste recycling techniques, though the recycling rate is lower [5, 6]. Recent advances in recycling entail innovative solutions to unlock socioeconomic opportunities, looking at waste as generating income [5, 6, 19]. Furthermore, these countries have started recovery policies and regulations to reduce unsustainable waste disposal—for example, Rwanda and South Africa [19, 20, 34].

Rwanda has evolved as one of the African countries pioneering in waste management and recycling, taking a practical system in which the environment and climate change are at the core of all policies, programs, strategies, and plans as realized in her vision 2012, 2020–2050 thus championing circular economy [34–36].

Rwanda has the most significant Green Fund (FONERWA) in Africa, one of the first national environment and climate change investment funds in Africa that advances innovation investments in public and private projects that initiate transformative change in Rwanda, for example, recycling [37, 38].

Rwanda has done well to mitigate plastic pollution and was the first East African Country to ban single plastic use in 2008 and revised in 2019 [20, 39]. Plastic waste is a severe global environmental challenge; only 15% is recycled. It negatively impacts the environment and affects plants, animals, and people. Plastic takes long to degrade, and its effects are prolonged. Thus, plastic recycling is a remedy to address the problem [20, 36, 40]

Plastic recycling in Rwanda is a collaborative engagement between the Government, private sector, international agencies, and local actors [40]. To effectively enhance plastic recycling, the Government of Rwanda signed the New Plastics Economy Global Commitment Report 17 with European Union (EU) to eliminate endless packaging and stimulate demand for recycled plastics. Recycling infrastructure is in place and reinforced by innovative technologies (**Figure 1**) [23, 40].

Rwanda has invested in electronic waste management and recycling waste from electronics like computers and mobile phones. E-waste management is challenging in Africa, has hazardous materials, accelerating severe impacts on human health and the environment. It contributes to 5% of greenhouse emissions [41, 42]. This is worsened by a lack of knowledge, absence of environmental policies, and insufficient funding; thus, e-waste is disposed of indiscriminately in open grounds and dumpsites (**Figure 2**) [40, 43].

Rwanda's electronic waste policy and regulatory framework made recent advances in electronic waste infrastructure high-tech and involve dismantling and recycling facilities, able to process up to 10,000 metric tons of e-waste annually, thus championing a circular economy in Africa [41, 44]. However, this sector is unexploited, and it is now starting to take off, and a lot has to be done in terms of research, policies, financial management, and funding [39, 40].

Africa has the potential for waste-to-energy, producing power from waste material. Recycling energy from waste is challenging and requires adequate waste management infrastructure, including the 3Rs; reducing, reusing, and recycling [38].

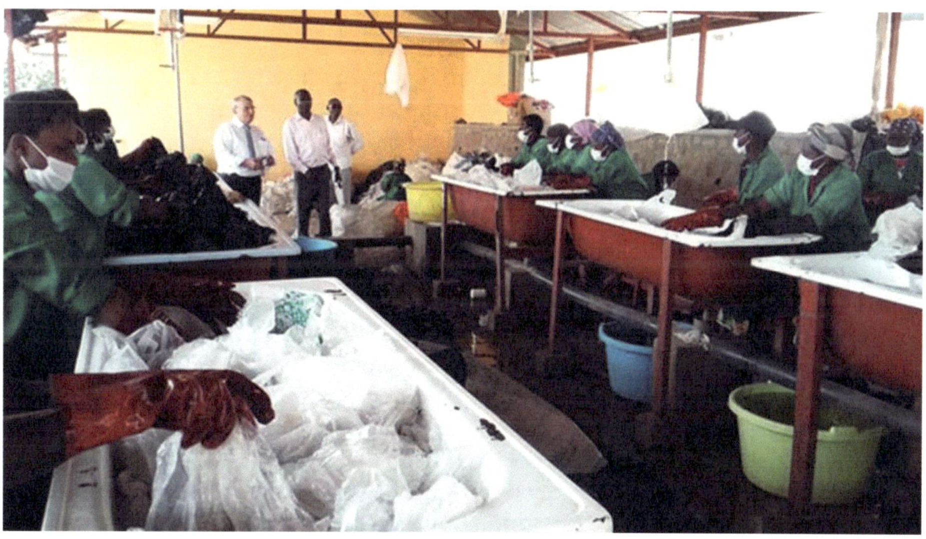

**Figure 1.**
*How recycling is done in a recycling facility in Rwanda (Source: The East African, Johnson Kanamugire, January 21st, 2018).*

**Figure 2.**
*E-waste íecycling in Rwanda (Souíce: ľiade foí Development News by EIĬ, Michelle Kovacevic, July 06, 2020).*

Waste to energy plants is expensive to build and is capital intensive. Most African countries cannot afford it [41, 44].

Rwanda has taken initiatives to recycle waste to produce energy in partnership with international investors though the focus is on biogas digesters to convert energy using organic waste [37, 41].

## 2.2 Recycling in South Africa

South Africa has made great efforts in managing waste management by enacting proactive policies, the white paper of 2000 and the National environmental management Act of 2008 established the integrated solid waste management system, the focus is on waste minimization, increasing collection coverage, and reducing waste taken to landfills [15, 45].

South Africa is among the top recycling countries in the world. South Africa's recycling industry is robust and economically oriented with a structured institutional and policy framework, focusing on recycling to protect the environment and human health [2, 17, 33]. This is reinforced by the National Environmental Management Act (NEMA) of 1998 and the National Waste Management Strategy (NWMS) of 2016, which propagates waste prevention through 3Rs, reduce, reuse and recycle; and responsible stewardship for the environment [15, 45].

South Africa's waste management is sustainable, with higher collection coverage of household waste at 59% in its metropolitan cities, Cape Town, Johannesburg, and Pretoria. Sustainable waste management is vital in recycling, and developed countries like Germany, with a higher recycling rate, can attest to that [7, 9].

South Africa's recycling industries recycle paper, steel, bottles, plastics, and energy being environmental champions in Africa in those areas. However, recycling is done from landfills, and informal waste pickers are the backbone of this activity where they retrieved recyclable resources, thus supporting formal waste management systems. This is done at all disposal streams; streets, landfills, open dumps, municipal trucks, or by itinerant consumers (**Figure 3**) [46, 47].

South Africa are champion in Africa's circular plastic industry, with an input recycling rate of 43.7%, above that of Europe's recycling rate, which presently is at 31.1% [24, 45]. The plastics industry is well organized under plastics SA with a mandate to

**Figure 3.**
*Plastic Recycling in South Africa (Source: Recycling International, Kirstin Linnenkoper, August 12, 2019).*

foster a vibrant and sustainable industry. The organization takes care of recycling, resource consumption, and energy recovery [48].

However, plastic recycling has challenges, ranging from mixed plastics streams like plastic trays used in food packaging and hard-to-remove remains, which are very difficult to recycle because of a lack of equipment and is capital intensive. The recycled product cannot pay this off. However, developed countries can recycle these plastics for oil or energy using available incinerator technologies, which explicates the gap in recycling [45, 46].

South Africa's metal packaging recycling is vibrant, with a recycling rate of 75%, a global champion in metal packaging recycling. Recycled metal packaging includes food tins, tin foil, and aluminum beverage cans [33, 49].

The extended producer responsibility (EPR) fund introduced by the South Africa Government has helped accelerates recycling in the paper and packaging industry and generates employment. The industry also invests in research to enhance the development of alternative usage for recycled paper and improve its quality [18, 21, 50].

South Africa recycles bottles through returnable Bottle Systems, where glass bottles are returned to the beverage producer for hygiene cleaning, its inspected and refilled. This system is one of the well-organized in the world [45, 48]. Glass is recycled using a system of closed-loop, recycling material from the same product, and the recycling rate of glass is at 40% [48, 50].

Despite being an environmental giant in recycling, South Arica faces challenges of access to quality and clean materials before they are retrieved at the landfills. This has been worsened by laxity on the regulation to separate waste at the source, making the process expensive, explicates the gap in recycling in Africa [22, 49, 50].

### 2.3 Challenges of recycling in Africa

Africa continent has done well in the legislative structure, 60% has vibrant policies, and regulations on waste management which is the framework in which recycling can ensue but they are geared toward waste disposal to dumpsites and landfills [18, 47].

The continent continues to face the challenges of unsustainable waste management since the sector is given less political priority and waste management systems have resulted into waste being seen as of no value that it can provide to local economies, impacting on environment and public health [18, 48].

Consequently, the African continent has not fully exploited the potential of waste recycling and the benefits derived from recycling in terms of economic and environmental, this is a major gap in recycling and challenges faced [45, 46].

Recycling is done by informal waste pickers, who collect 90% of recyclables in the continent cities, on average, they assemble between 20% and 50% of generated waste and form the waste management supply chain, collecting, sorting and selling discarded waste [46, 51]. However, their activities remain in the periphery of waste management system, no targeted incentives to boast their contribution in the waste management landscape.

They face social stigma, harassment, and abuses from municipal and security officials, impacting negatively on their operations [25, 46, 51]. There is an urgent need for this group to be integrated into the waste management landscape, as stakeholders by being formally organized into community-based organizations (CBO) and self-help groups, thus enhancing their contribution, increasing employment opportunities and sustainable waste management, and key to unlocking these opportunities [25, 27].

There are more opportunities that can be created through recycling organic waste in Africa which forms 70 to 80%, and plastics 13%. However, this area has been overlooked, the continent should take a paradigm shift, and focus on organic recycling [15]. Recycling organic waste will offer more economic and social opportunities, it is not capital intensive and more accessible, for instance, recycling agricultural waste for fuel briquettes, biogas generates clean cooking energy to charcoal use, thus reducing deforestation and may help minimize indoor pollution, environmental degradation, which contributes to greenhouse gas emissions, averting the effects of climate change [27].

The African continent can explore formal plastic recycling, using environmentally friendly technologies to treat that waste [42, 48]. It should involve public-private partnership, incentives be given to boast their confidence for increased investments in this sector, creating employment opportunities and improving livelihoods [48]. Africa's major economies, Nigeria, Morocco, and South Africa, have taken up plastic recycling. South Africa's plastic recycling has created over 4000 and 30000 jobs, both direct and indirect, and added an additional $450 million into the economy [42, 45, 48].

Projections, environmentally shows that to meet global developmental supplies into the future, require a 200% increase in natural resource removal by 2050, a big challenge where people have overexploited the natural resources, including Africa, through environmental degradation [11, 22]. This can be addressed through increased recycling to reduce the extortion of new resources for each production cycle and Africa should take those initiatives. Reforms must be made to make new legislation and reinforce weaker ones [18, 51].

## 3. Conclusion

The future for Africa's waste management is vibrant and the opportunities to increase recycling as a sustainable avenue in waste management are yet to be seized. Based on the fact the majority of waste generated is recyclable, the adoption of technology and building capacities in waste management gives hope to better environments and improved health of people. Moreover, the economic pillar of waste management through recycling forecasts the numerous jobs and wealth creation avenues that the sector can bring about. The mindset shift required to adopting and scaling up recycling as a viable option still requires more commitment from every sector for sustainable development to be a reality. Africa's potential in recycling has not been fully realized portending the numerous untapped opportunities.

Further research can be focused on the following:

1. Dissemination of technologies in Africa for Sustainable Waste Management through Recycling.

2. Assessment of Governance Structures for Adoption of Efficient Waste Management through Recovery and Recycling.

3. Examine the socio-cultural attributes that enhance recycling in Africa's context.

## Author details

Florence Akinyi Ogutu* and Bessy Kathambi
Department of Earth and Climate Sciences, Wangari Maathai Institute of Peace and Environmental Studies, University of Nairobi, Nairobi, Kenya

*Address all correspondence to: florence.akinyi@gmail.com

## References

[1] Ferronato N, Torretta V. Waste mismanagement in developing countries: A review of global issues. IJERPH. 2019;**16**:1060

[2] Bello IA, Ismail MNB, Kabbashi NA. Solid Waste Management in Africa: A Review. International Journal of Waste Resources. 2016;**6**:216. DOI: 10.4172/2252-5211.1000216

[3] Bundhoo ZMA. Solid waste management in least developed countries: Current status and challenges faced. Journal of Material Cycles and Waste Management. 2018;**20**:1867-1877

[4] World Bank. Overview. 2020. Available from: https://www.worldbank.org/en/country/kenya/overview. [Accessed: 17 August 2020]

[5] Asante KA, Amoyaw-Osei Y, Agusa T. E-waste recycling in Africa: Risks and opportunities. Current Opinion in Green and Sustainable Chemistry. 2019;**18**:109-117

[6] Perez TS. The discursive power of recycling: Valuing plastic waste in Cape Town. Worldwide Waste: Journal of Interdisciplinary Studies. 2021;**4**:8

[7] Ferronato N, Ragazzi M, Gorritty Portillo MA, et al. How to improve recycling rate in developing big cities: An integrated approach for assessing municipal solid waste collection and treatment scenarios. Environment and Development. 2019;**29**:94-110

[8] United Nations Human Settlements Programme. Solid Waste Management in the World's Cities: Water and Sanitation in the World's Cities. 2010. UN-HABITAT/Earthscan; 2010. p. 228

[9] Abubakar IR. Governance challenges in African Urban Fantasies. In: Home R, editor. Land Issues for Urban Governance in Sub-Saharan Africa. Cham: Springer International Publishing; pp. 155-169

[10] Darhamsyah. Environmental Governance Urban: Public Participation and Sustainable Development. Epub ahead of print 6 April 2019. DOI: 10.5281/ZENODO.2633632.

[11] Cobbinah PB, Erdiaw-Kwasie MO, Amoateng P. Rethinking sustainable development within the framework of poverty and urbanisation in developing countries. Environment and Development. 2015;**13**:18-32

[12] Karak T, Bhagat RM, Bhattacharyya P. Municipal solid waste generation, composition, and management: The world scenario. Critical Reviews in Environmental Science and Technology. 2012;**42**:1509-1630

[13] Shahmoradi B. Collection of municipal solid waste in developing countries. International Journal of Environmental Studies. 2013;**70**:1013-1014

[14] Godfrey L, Tawfic Ahmed M, Giday Gebremedhin K, et al. Solid waste management in Africa: Governance failure or development opportunity? In: Regional Development in Africa. London, UK: IntechOpen; 2019

[15] Godfrey L, Tawfic Ahmed M, Giday Gebremedhin K, et al. Solid waste management in Africa: Governance failure or development opportunity? In: Edomah N, editor. Regional Development in Africa. London, UK: IntechOpen; 2020

[16] Gilbert N, Ziqiang Y, Hongzhi M. Current situation of solid waste

management in East African countries and the proposal for sustainable management. African Journal of Environmental Science and Technology. 2021;**15**:1-15

[17] Kathambi BE, Ogutu FA. Ecological impacts of improper waste management frameworks on biodiversity conservation in Nairobi County, Kenya. Journal of Biodiversity and Environmental Sciences. 2021;**19**:91-101

[18] World Bank. What a waste: An updated look into the future of solid waste management. 2018. Available from: worldbank.org/en/news/immersive-story/2018/09/20 what-a waste-an-updated look-into the future-of- solid-waste-management

[19] Magrini C, D'Addato F, Bonoli A. Municipal solid waste prevention: A review of market-based instruments in six European Union countries. Waste Management & Research. 2020;**38**:3-22

[20] Ogutu FA, Kimata DM, Kweyu RM. Partnerships for sustainable cities as options for improving solid waste management in Nairobi city. Waste Management & Research. 2021;**39**:25-31

[21] Geissdoerfer M, Savaget P, Bocken NMP, et al. The circular economy—A new sustainability paradigm? Journal of Cleaner Production. 2017;**143**:757-768

[22] International Solid Waste Association (ISWA). Annual review report: Working together towards a cleaner, healthier planet. International Solid Waste Association (ISWA). 2017. Available from: https://www.lswa.org/fileadmin/galleries/publications ISWA Reports/

[23] UNEP. New Plastics Economy Global Commitment Circular Economy in the Africa-EU Cooperation. Nairobi: 30 Ellen MacArthur Foundation & UNEP. 2019. Available from: https://ellenmacarthurfoundation.org

[24] UNEP. Africa Waste Management Outlook. Nairobi: UNEP; 2018

[25] UNDP. Sustainable urbanization strategy, support to sustainable, inclusive, and resilient cities in the developing world. UNDP. 2016

[26] Angoua ELE, Dongo K, Templeton MR, et al. Barriers to access improved water and sanitation in poor peri-urban settlements of Abidjan, Côte d'Ivoire. PLoS ONE. 2018;**13**:e0202928

[27] Ogutu FA, Kimata D, Kweyu R. Factors affecting the use of environmental values and ethics in solid waste management in Nairobi city county. International Journal of Environmental and Health Sciences. **1**:1-9

[28] Aina YA, Wafer A, Ahmed F, et al. Top-down sustainable urban development? Urban governance transformation in Saudi Arabia. Cities. 2019;**90**:272-281

[29] Bottero M, Caprioli C, Cotella G, et al. Sustainable cities: A reflection on potentialities and limits based on existing eco-districts in Europe. Sustainability. 2019;**11**:5794

[30] Eisenmenger N, Pichler M, Krenmayr N, et al. The sustainable development goals prioritize economic growth over sustainable resource use: A critical reflection on the SDGs from a socio-ecological perspective. Sustainability Science. 2020;**15**:1101-1110

[31] Hansson S, Arfvidsson H, Simon D. Governance for sustainable urban development: The double function of SDG indicators. Area Development and Policy. 2019;**4**:217-235

[32] Kathambi BE, M'Ikiugu HM. Implementing sustainable development Goals 1, 3.9, 7, and 13 through adoption

of green concept in environmental management: Case of Nairobi, Kenya. Journal of Biological and Environmental Sciences. 2018;**12**:1-10

[33] Zuckerman A. 60 Recycling Statistics: 2020/2021 Data, Trends & Predictions. CompareCamp. 2020. Available from: https://comparecamp.com/recycling-statistics/

[34] Irbagza. Environment protection and waste management in Rwanda. New Time Rwanda. 2021. Avaialble from: https://www.newtimes.co.rw › opinions › environment

[35] Shamim A, Mursheda A, Rafiq I. E-Waste trading impact on public health and ecosystem services in developing countries. International Journal of Waste Resources. 2016;**5**:1-12. DOI: 10.4172/2252-5211.1000188

[36] Karenzo ME. Waste_Management_in_Kigali:City. Rwanda. 2021. Available from: https://www.academia.edu›50631492›economic,socialsciences,Agriculturee

[37] Government of Rwanda. Rwanda Green Fund. Rwanda: Ministry of Environment; 2019

[38] Government of Rwanda. Rwanda's National e-waste Management Policy. Government of Rwanda, E-waste. 2019. Available from: https://iins.org›rwandas-national-e- waste-management-policy

[39] Government of Rwanda. Rwanda Vision 2020. Rwanda: Government of Rwanda. 2019. Available from: https://rsb.gov.rw›...›files›Vision_2020_Booklet.pdf

[40] Danielsson M. The Plastic Bag Ban in Rwanda: Local Procedures and Successful Outcomes: A Case Study on how Rwanda Implemented a Nation-wide Ban on Plastic Bags. Uppsala University; 2017

[41] Daniel R. E – Waste, Marketplace Africa: The Rising e-waste crisis is being Reckoned with in Rwanda, one gadget at a time. CNN. 2021

[42] Blaauw D, Pretorius SR. The economics of urban waste picking in Pretoria. African Review Economic Finance. 2019;**11**:129-164

[43] Republic of Rwanda. City of Kigali, Police Launch Joint Security, Hygiene Drive. Kigali, Rwanda: Republic of Rwanda; 2018

[44] Kyere VN. Environmental and Health Impacts of Informal E-waste Recycling in Agbogbloshie. Accra, Ghana: Recommendations for Sustainable Management; 2016

[45] Government of South Africa. SA Plastics Pact. South Africa: Government of South Africa; 2020

[46] Godfrey L. Quantifying economic activity in the informal recycling sector in South Africa. South African Journal of Science. 2021;**117**(9/10). DOI: 10.17159/sajs.2021/8921

[47] Research Markets. South Africa Waste and Scrap Recycling Industry Report 2021. South Africa: Research Markets; 2021

[48] SAPPRO. Plastics-Recycling-in-SA-2019. South Africa: SAPPRO; 2019

[49] Yasmin M. Recycling of Waste and Scrap 2019. South Africa: Research Markets; 2019

[50] UNEP. From Pollution to Solution: A global assessment of marine litter and plastic pollution Synthesis. 2021. Available from: chrome-extension://efaidnbmnnnibpcajpcglclefindmkaj/https://wedocs.unep.org

[51] UNEP. Global Waste Management Outlook Report in Africa. UNEP: Nairobi; 2019

Chapter 6

# Recycling Asphalt Pavements: The State of Practice

*Hesham Ali and Aditia Rojali*

## Abstract

Millions of miles of paved roads in the world need rehab, maintenance, and upkeep. To sustain this effort, pavement engineers are challenged to find innovative ways to maintain the roads in an economical and environmentally responsible fashion. Pavement recycling is a clear way to achieve sustainability. This chapter provides a brief introduction and outlines the state of practice with respect to asphalt pavement recycling. The reuse of various pavement layers (asphalt, asphaltic base, granular base, and subbase) is discussed. The ways of utilizing the material, from the highest to the lowest economical use, are presented. The methods of recycling individual layers or combining multiple layers are discussed.

**Keywords:** reclaimed asphalt pavement, recycling asphalt, rejuvenator, base course, subbase, embankment, emulsion, foamed asphalt, full depth reclamation

## 1. Introduction

Removed pavement material through the process of reconstruction or resurfacing work is known as Reclaimed Asphalt Pavement (RAP). It is recognized to contain usable aggregate and asphalt binder. Historically, asphalt mix with RAP began in 1915. Its usage is increasing since the rise in oil prices happened in 1970s. RAP became an attractive option to reduce economic cost of roadway construction. This is due to decreasing demand for asphalt binders. State DOTs used RAP for many years before the beginning of Superpave era in the late 1990s. The maximum quantity of RAP is found in surface layer. In the US, the average RAP content in asphalt mixture is 12–15% for a new construction project [1]. RAP application now is accepted all over the world [2]. Denmark, France, Sweden, and Germany were found to be the significant RAP users in Europe. In addition to these four countries, the Netherlands uses 30–40% proportion of old asphalt for the new mixture [2]. Recently, pavement research community is working on advancing RAP usage practice by raising RAP content to 40-50% and up to 100% [3].

Roadways in the US comprise more than 90% built with asphalt layer. The pressure needs to maintain the performance and rehabilitate an aging asphalt is growing and will continue to occur. The same material used to construct the roadway can be reused to prolong pavement life cycle. RAP usage will be projected to be more common as RAP offers more environmental benefits than virgin material. A primary activity of pavement maintenance is milling operation. Milling process produced RAP that consists of aged asphalt binder and original aggregates. RAP usage is aligned with global Sustainable Development goals because of provident use of natural resources. RAP application promotes natural resource preservation and reduces landfill occupation from disposal of roadway construction and maintenance projects. Energy consumption will also be reduced since there will be less transportation demand for virgin material from site plant or material disposal to landfill site. In 2017, 76.2 million tons of RAP were recycled in the US and it is increased to 89.2 million tons in 2019 [4]. In EAPA's (European Asphalt Pavement Association) report released in 2020, 50 million tons or RAP were recycled in all European Union countries [5]. The report from 2016 showed RAP usage for over 60 million tons in China [6].

Economic savings have been highlighted by numerous literatures through different type of analysis [7–10]. Silva et al. [10] reported cost savings of $300 million when hot mix asphalt contained 30 million tons of RAP. RAP usage positively impacts the environment as RAP application reduces waste and the inherent issues around the disposal of highway materials [11, 12]. A cost saving from $6.80 to $3.40 per ton will be generated when a mixture contain RAP from 15% to 40% [13]. A predicted reduction of energy consumption from 16% to 25% for a mixture that uses 16% to 25% RAP [14].

One of the significant value-added measures available to highway agencies is increasing the application of RAP in the construction and rehabilitation of asphalt pavements. RAP used in new pavement mixtures replaces a portion of the expensive virgin binder and virgin aggregate and provides protection from the volatility of asphalt binder price variability. The cost to prepare RAP for use in new mixtures is consistent. The value of the RAP is essentially the cost per ton of virgin mix (i.e., value of the virgin aggregate and virgin asphalt) minus the cost to acquire and process the material for use in new mixtures. It is suggested a mix with 30% RAP has equivalent performance to a mix with all virgin material [15].

Experiment with Hamburg Wheel tracking test that a mixture with 30% and 50% combined with a soft binder showed similar performance with virgin aggregate mixture [16]. The aged, stiff binder can be compensated by the application of rejuvenating agents, softer virgin binder grade, and therefore produce acceptable pavement performance [3].

Although high RAP content is theoretically possible, advanced processing of RAP is required to make high content of RAP produce acceptable performance. Advance RAP processing like warm asphalt technology and testing will increase project costs. For this reason, advanced processing could inhibit the potential growth of high RAP content.

Highway agencies in the US have specifications for RAP application and most of them limit the amount of RAP used in specific asphalt or mixture types. In a 2009 survey conducted on behalf of the Federal Highway Administration (FHWA) and the American Association of State Highway and Transportation Officials (AASHTO), one of the major concerns cited by State Highway Agencies (SHAs) regarding limited RAP application is the quality of the blended virgin and RAP binder qualities. For a high RAP mixture, the quality concern is even more emphasized.According to AASHTO M 323, the assumption which is placed to formulate current binder selection guidelines for RAP mixtures is that the virgin and RAP binders are blended effectively. Based on these guidelines, some State highway agencies give full credit to the contribution of the RAP binder to the asphalt content, which may result in an under-asphalted mixture if complete or significant blending is not occurring. A survey conducted in the US by NCHRP Synthesis 495 indicated that 100% of the RAP binder was available in the mix from 77% of all respondents. However, the study found only 42% of all respondents consider 100% of the RAS binder available in the new mix [17]. The past studies showed that the virgin and RAP binder blending amount is in the middle of complete blending and no blending (i.e., the RAP aggregate behaves like a black rock). However, there is no consensus on how to accurately determine the amount of blending with a direct method.

RAP was utilized for multiple purposes in pavement engineering, from asphalt concrete, to unbound granular layer, to embankment fill. Asphalt binder is the costliest component in the asphalt mixture. As the aged asphalt contained in RAP reduces the needed binder for the new mix; therefore, the most valuable usage of RAP is in asphalt mixtures. Similar performance of road section containing at least 30% asphalt performs similarly to virgin section [18].

A summary of reclaimed material usage in pavement engineering is summarized in **Table 1** [19, 20].

## 2. Recycling the asphalt concrete layer

This section describes the recycling of the asphalt layer alone. Despite the various application of RAP content in pavement asphalt layers, there is an increasing trend on application of a high RAP proportion in the new mixture [21]. Moreover, Perez and Maicelo [21] discussed from various literature the performance, fatigue life, rutting, and cracking of RAP with different proportions and mixing techniques in asphalt mixtures.

Naulkha [22] summarized the mechanical properties and performances of RAP. RAP creates a rut and skid-resistant layer due to an aggregate interlocking effect. Air voids are reduced with more RAP in the mixture.

Asphalt recycling methods are classified based on treatment place, temperature treatment, characteristic material, type of binder, etc. In general, there are five categories of asphalt recycling, and these methods can be combined one with another [23]:

*Cold Planning*: Cold planning (milling) uses specific equipment to remove asphalt layer into specified depth, longitudinal profile, and cross-slope.

*Hot Recycling*: A process to produce recycled mix from the combination of RAP with virgin aggregate, new binder, and rejuvenating agents, as needed to be conducted in central plant.

*Hot In-Place Recycling*: The process consists of heating and softening the existing asphalt, milling it to a specified depth, and then mixing and placing it with conventional paving equipment. All recycling processes are done on-site. Treatment depths range from ¾, 2, and 3 inches. Hot in-place recycling subcategories are Surface Recycling (Resurfacing), Remixing, and Repaving.

*Cold Recycling*: Asphalt recycling process without heat application. Cold recycling can be done in-place and in-plant.

*Full Depth Reclamation*: Full thickness of asphalt layer and predetermined underlying materials is pulverized, blended, and then stabilized using various methods like mechanical stabilization, binder-based stabilization, and chemical stabilization.

| Application/use | Material |
|---|---|
| Asphalt Concrete—Aggregate (Hot Mix Asphalt) | Blast Furnace Slag<br>Coal Bottom Ash<br>Coal Boiler Slag<br>Foundry Sand<br>Mineral Processing Wastes<br>Municipal Solid Waste Combustor Ash<br>Nonferrous Slags<br>Reclaimed Asphalt Pavement<br>Roofing Shingle Scrap<br>Scrap Tires<br>Steel Slag<br>Waste Glass |
| Asphalt concrete—aggregate (cold mix asphalt) | Coal bottom ash Reclaimed asphalt pavement<br>Blast-furnace |
| Asphalt concrete—aggregate (seal coat or surface treatment) | Blast-furnace slag<br>Coal boiler slag<br>Steel slag |
| Asphalt concrete—mineral filler | Baghouse dust<br>Sludge ash<br>Cement kiln dust<br>Lime kiln dust<br>Coal fly ash |
| Asphalt concrete—asphalt cement modifier | Roofing shingle scrap<br>Scrap tires |
| Portland cement concrete—aggregate | Reclaimed concrete |
| Portland cement concrete—supplementary cementitious materials | Coal fly ash<br>Blast-furnace slag |
| Granular base | Blast-furnace slag<br>Coal bottom ash<br>Coal boiler slag<br>Mineral processing wastes<br>Municipal solid waste<br>Combustor ash<br>Nonferrous slags<br>Reclaimed asphalt pavement Reclaimed concrete<br>Steel slag<br>Waste glass |
| Embankment or fill | Coal fly ash<br>Mineral processing wastes<br>Nonferrous slags<br>Reclaimed asphalt pavement Reclaimed concrete<br>Scrap tires |
| Stabilized base—aggregate | Coal bottom ash<br>Coal boiler slag |
| Stabilized base—cementitious materials (Pozzolan, pozzolan activator, or self-cementing material) | Coal fly ash<br>Cement kiln dust<br>Lime kiln dust<br>Sulfate wastes |
| Flowable fill—aggregate<br>Flowable | Coal fly ash<br>Foundry sand<br>Quarry fine |

| Application/use | Material |
|---|---|
| Flowable fill—cementitious material (pozzolan, pozzolan activator, or self-cementing material) | Coal fly ash<br>Cement kiln dust<br>Lime kiln dust |

**Table 1.**
*Recycled materials usage in pavement construction [19, 20].*

## 2.1 Milling and reuse of RAP

Milling and full depth removal are the primary processes of asphalt pavement removal found in maintenance and rehabilitation project. In a single pass, up to a 2-in surface layer can be removed by a milling machine. Meanwhile, a full depth removal uses pneumatic pavement breakers to rip and break the pavement. The process will continue by pooling the removed material to a haul-trucks with a front-end loader and transporting the material to asphalt plant processing facility or use in place processing equipment. The RAP will be crushed, screened, and conveyed.

The most common RAP processing activity is conducted in central processing plants where a self-propelled pulverizing machine is used to pulverize RAP materials. The process includes mixing with stabilized base courses or granular material. Recent practice in hot in-place and cold in-place paving operations involves continuous train operations of partial depth removal, mixing RAP with virgin aggregate, binder, rejuvenating agents, and then placing and compacting in one pass.

### *2.1.1 Use of RAP in plant produced HMA (highest economic value)*

In-plant production HMA consists of mixing of new materials with RAP to produce hot mix asphalt. Recycling agents are utilized to improve the chemical composition, rheology, and dispersion of aged bitumen contained in RAP. Adding a recycling agent is proven to produce a comparable performance to a virgin binder. According to ASTM D4552 "Standard Practice for Classifying Hot-Mix Recycling Agents", recycling agents are categorized mainly into three, namely: asphalt cements, naphthenic aromatic oils, and paraffin oils. The dose of a rejuvenating agent for RAP can be specified by ASTM D4887 "Standard Practice for Preparation of Viscosity Blends for Hot Recycled Bituminous Materials". Absolute viscosity and performance properties are target parameters for asphalt mixes with RAP. Recycling agent's critical parameters in their physical and chemical properties play a controlling factor in the selection of recycling agents [24]. The primary functions of rejuvenating agents are to increase maltene fractions and to decrease asphaltene. However, there is no guarantee that the durability of restored aged asphalt will increase with high restoration capacity. RAP mixes with an appropriate composition have been shown to equal and even surpass the virgin material performance.

Several works have been conducted in the past to investigate mixture performance with RAP as summarized in [25]. The stiffness of the mixes will be increased with higher RAP contents. When a soft binder is used with RAP content of 0 to 30%, a very less significant effect was found [26]. Tarsi et al. [25] also showed from their review that lower resistance to fatigue of the mixes was found when the mix used higher RAP contents [17, 27–29]. Meanwhile, different results showed higher RAP contents lead to higher rutting resistance [10, 17, 27, 30]. The mix that is not stabilized with rejuvenating agents but has higher RAP content showed higher stiffness, and the RAP mix stabilized by a rejuvenating agent performed the best fatigue resistance [10]. Zaumanis et al. [31] found all RAP mixes have similar stiffness with virgin mix. Lee et al. [32] showed no noticeable disparity in mixes with different RAP content. In the moisture damage test, several RAP mixes failed to comply with the standard criteria; on the other hand, a better moisture resistance than virgin material was shown by RAP mixes [25].

#### *2.1.2 Use of RAP as a base material (2nd highest)*

Base and the lower subbase layer are typically found underneath the asphalt layer. These lower layers have the function to reduce and spread loads within the subgrade. As a foundation for overall layers, the quality of base and subbase layer is significant to pavement service life. Subbase materials are typically coarser-grained than the granular base. Both materials are intended to offer the pavement structure's requisite bearing capacity and drainage. They are necessary for the longevity of pavement performance.

A resilient modulus measurement from field and laboratory showed that RAP application for base and subbase layer exhibited equivalent strength to dense graded aggregate [19]. Another study used RAP and virgin material blend with various percentages of total weight mixed aggregate which resulted in a comparable performance to the virgin base material [27].

Collins [28], an NCHRP Synthesis of Highway Practice identified an application of RAP for the unbound base and subbase layer with grading identified as the limiting factor for use. RAP used in base materials is found in 13 states, RAP used in subbase material in four states, and two states reported to use RAP application as stabilized base and shoulder aggregate [28]. The overall performance of RAP contained base and subbase layer has been identified from satisfactory to excellent. 16 state DOTs permit the use of 100% RAP as an aggregate unbound layer and five state DOTs limit it to 50% or less [30]. Potential cost benefit when RAP is used as an unbound layer with a 50% blend with virgin aggregate is approximately 30% [30].

An example of testing application for coarse aggregate is Micro-Deval test loss. Hoppe et al. [30] highlighted a test for Virginia typical coarse aggregates [33]. A test loss result of less than 15% were suitable for all applications including HMA layer with a more stringent bound layer requirement. In Ontario, Canada, a Micro-Deval test loss limits is specified as 21% for open-graded aggregate, and 30% for subbase material [33]. It is worth to mention that Ontario has high moisture effect than Virginia. Therefore, the Micro-Deval test can be applied for testing RAP application for the unbound layer.

The physical properties of RAP were found similar to crushed limestone. Hoppe et al. [30] also highlighted the gradation parameter of RAP material. RAP material has an equivalent gradation with crushed virgin aggregate. However, milling and stockpiling activities play an important role that may determine the proportion of fines in RAP, which typically, can be a large range [30]. Particles of original coarse aggregate are assumed to be strong and resistant to deformation. In contrast, accumulations of fine aggregate and asphalt mastic may be fragile or flexible. This flexibility is related to the asphalt condition like age, oxidation, and temperature [30].

Different findings on stiffness parameter in RAP specimens have been reported. Several studies have demonstrated that RAP has a relatively high modulus of elasticity, accompanied by considerable permanent deformations. In a comparison between 100% RAP specimens with dense graded aggregate, RAP specimens exhibit higher stiffness, resilient modulus value but lower shear strength. 100% of RAP specimens showed largest permanent strain regardless of high resilient modulus value. When load applies to asphalt binder, it will cause progressive decomposition of the binder and may be the cause of the significant permanent strain [34]. RAP specimens tested as unbound aggregates showed high resilient modulus and larger permanent deformation. Unbound RAP mixture also showed temperature and viscous properties dependent [33]. When RAP increases, shear strength decrease and results in lower bearing capacity compared with virgin aggregate [35, 36].

Another work reported decreasing CBR value when using RAP in the subbase layer. Under loading application, RAP aggregate slides one to another [30]. RAP contained granular mixture showed noticeable amount of creep [37]. Cosentino et al. [33] suggested a minimum 75% standard aggregate mix with an non stabilized RAP for base layer application. It is also found that 25% granular RAP and 75% of limerock satisfy base layer requirement without stabilizing agent and 50% granular RAP with stabilizing agent [38].

Meanwhile, 100% RAP is not recommended as this will not produce acceptable base performance [33, 39]. This is due to decreasing shear strength as increasing RAP content. It is recommended that only 25% of RAP content mixed with virgin aggregate and it has to be conducted in the mixing plant [40]. It is found that onsite blending has unsatisfactory performance [39].

In Florida, it is reported from a field experiment with 100% RAP in the base layer showed acceptable performance, equivalent to base layer of limerock. It is recommended to use 0.12 to 0.15 for structural coefficient for use with AASHTO empirical equation and limerock bearing ratio test is not suitable for RAP aggregate [41].

To characterize the field performance of RAP mixtures, the following tests were found to produce statistically significant performance indicators of recycled aggregates in unbound pavement layers [42]:

- Screening tests for sieve analysis and the moisture-density relationship;
- Micro-Deval for toughness;
- Resilient modulus for stiffness;
- Static triaxial and repeated load at optimum moisture content and saturated condition for shear strength; and
- Frost susceptibility (tube suction).

### *2.1.3 Use of RAP in embankment (lowest economic value)*

Stockpiled RAP material can be used as a fill for embankment and backfill construction and as a practical alternative to stockpiled materials [38]. In this case, RAP materials are treated as granular fill. RAP material may be mixed with soil and/or fine graded aggregate when used as fill material or embankment. This option is most suitable when there is a market for reuse products or when RAP is not usable in asphalt concrete layer. FHWA reported that there are at least nine state DOTs that allow RAP material to be used as embankment or fill material. These include Connecticut, Indiana, New York, Tennessee, Kansas, California, Illinois, and Louisiana. RAP material performance for embankment or fill material is considered satisfactory and good [43]. RAP material for embankment requires minimal processing like primary crushing. A study in Wisconsin investigated the construction of embankments containing bituminous materials. The study recommended that construction should be conducted during summer to generate thermal preloading and reduce long-term settlement [40].

#### *2.1.4 Use of RAP in pavement-preservation treatments*

RAP uses in preservation treatments like micro surfacing, and cheap seals are gaining more popularity for a variety of reasons, similar performance to virgin materials, cost saving, and reduced environmental impact.

In California, despite increases in permitted RAP percentages in flexible pavement construction in California, there is still an overabundance of RAP in metropolitan areas. To cost-effectively employ recycled materials and reduce otherwise rising RAP stockpiles, agencies turned to chip sealing, slurry sealing, and microsurfacing [44].

Reclaimed asphalt pavement aggregate using polymer-modified emulsion is reported to perform well. A chip seal containing RAP and PG 76-22 rubber modified asphalt binder also performs positively. RAP slurry is widely used in Southern California [44]. RAP aggregate has lower absorption for emulsion than the virgin aggregate. RAP aggregate has also captured the emulsion instead of forming a mechanical bond as in a combination of virgin aggregate and polymer-modified slurry. In conclusion, the agencies in California will improve the specifications, particularly in treatment constructability, and increase economic and environmental benefits over the next few years based on performance evaluation.

A series of tests were conducted to compare a chip seal performance with RAP and virgin aggregate. Summary of the results is as follows:

Lifecycle cost analysis showed that a cheap seal with RAP cost lesser, around 23–37%, than a cheap seal with virgin aggregate but exhibits similar performance [45].

- Sand patch test: The mean texture depth was higher than chip seals with virgin aggregate.
- Skid resistance test: No significant difference between cheap seal with RAP and virgin aggregate.
- Direct shear test: No significant difference between cheap seal with RAP and virgin aggregate.

### 2.2 In-place recycling of asphalt layer

In-place recycling is gaining a position as a preferred method due to the demand for rehabilitation of old pavements exceeding the development of new roads. The attractiveness of in-place recycles is also caused by its ability to address backlog in pavement rehabilitation [46].

The cold in-place process utilizes a "train" of equipment such as tanker trucks, milling machines, crushing, and screening, mixers, pavers, and rollers. The common method of recycling HMA pavement is to grind the top layer of the old

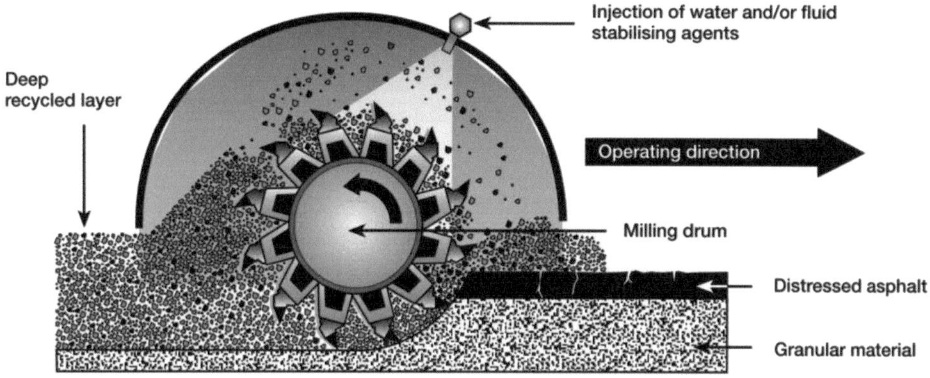


**Figure 1.**
*Process inside the milling machine for typical cold in place recycling [46].*

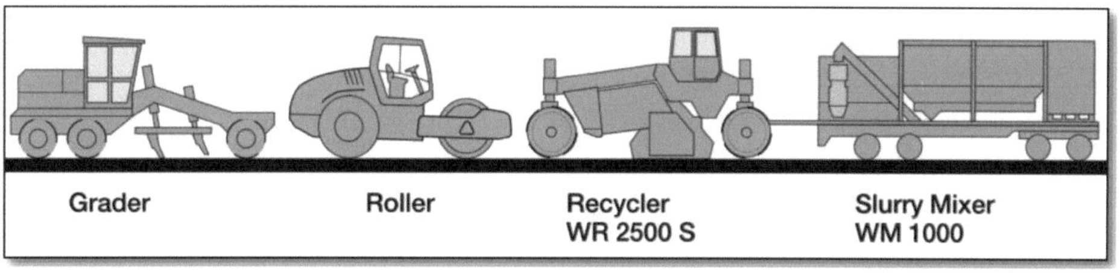


**Figure 2.**
*Recycling train applying cement slurry [42].*

**Figure 3.**
*A schematic of a recycling train using emulsion slurry [42].*

pavement. The process continues with transporting the RAP to the asphalt plant and stockpiling the RAP. RAP in stockpile will be incorporated back into new HMA. In-place recycling offers a more cost-efficient method. In-place method eliminates the trucking and handling of the recycled asphalt by performing the complete process in one pass. In place method consist of hot in-place and cold in-place recycling.

Cold in-place recycling mainly occurs inside the milling chamber. Milling and mixing the recycled material with stabilizing agents like cement, bitumen emulsion, and slurry water combined in one single operation as shown in **Figure 1**. Meanwhile, the recycling train will have a different configuration that depends on the types of recycling application and stabilizing agent. Included in the train, there will be recycling machine as the locomotive that pushes or pull the embedded equipment.

**Figure 2** shows a typical train for recycling with cement slurry stabilizing agent. The slurry will be pumped to the recycler with application rate metered before mixing (**Figure 3**).

### *2.2.1 Hot in place recycling*

The rehabilitation process of old HMA pavement onsite, conducted in one operation, is called hot in-place recycling. The hot in-place recycling process starts with heating the asphalt pavement to a temperature that eases the process of milling. The process continues with RAP improvement by including virgin aggregate, and rejuvenator to the hot millings. Furthermore, the paver machine spreads and compacts the RAP. The type of processes hot in-place recycling are surface recycling, remixing, and repaving [23].

Surface recycling is the simplest one and can only rehabilitate the uppermost layer of pavement. Surface recycling starts with heating, continued by milling and placing the RAP without further processing except for rejuvenating agent that may be needed. Surface recycling is typically continued by a bituminous surface treatment overlay. The remixing process, on the other hand, allows recycling depth of 2 inches or more due to multiple passes of heating and scarifying. The remixing process also involves virgin aggregate and binder addition to the hot millings. Repaving places new HMA layer overlay above the hot in-place recycled lift. Hot in-place recycling was found to have less initial project cost and reduced length of lane closure as compared with conventional hot mix asphalt. Several other benefits of hot in-place recycling are as follows [23]:

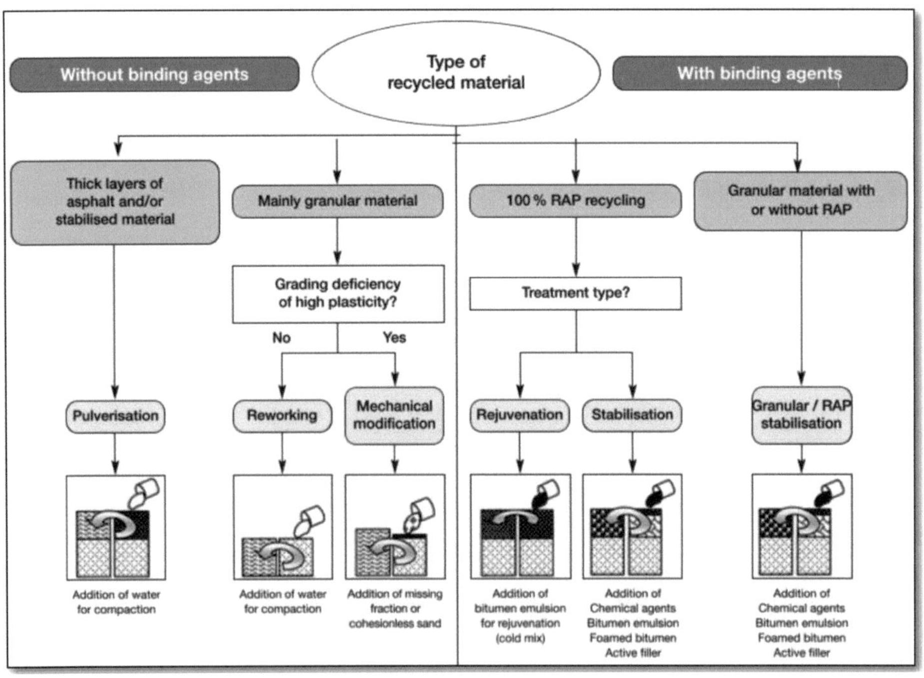


**Figure 4.**
*Wirtgen's framework of cold pavement recycling [46].*

- Recycling rate can achieve up to 100%.
- Less energy is used compared to the other rehabilitation methods.
- The amount of aggregate breakage is reduced due to heating and softening of the old asphalt pavement.
- Reduced transportation needed to move new material and stockpiling RAP.

#### *2.2.2 Cold in place recycling*

Cold in-place recycling is a method of aging pavement rehabilitation conducted at the project site. Therefore, Cold in place can reduce the need to loading and transporting material. The process includes milling and crushing the aging asphalt concrete, rejuvenating, and then laying and compacting the RAP. Asphalt pavements rehabilitated with cold in-place method were found to perform well [47]. Cold in-place is the recommended method for rehabilitating low-volume roads that have AADT of less than 2000, in Iowa, United States. Cold in-place roads have, on average, predicted service life from 15 to 26 years (**Figure 4**) [47].

## 3. Full depth reclamation: recycling and combining the asphalt Layer Base and subbase layers in-place

A portion or full pavement layer material from asphalt to subgrade is recycled through grinding and blending process to produce homogenous material. Typically, the blending product sufficiently performs as base material for new asphalt concrete layer even without stabilizing agents. However, project evaluation is required to assess the necessity of involving stabilization. Common methods of stabilization are mechanical, chemical, and bituminous. Mechanical stabilization uses granular materials such as virgin aggregate or RAP or crushed Portland cement concrete. Chemical stabilization uses lime, Portland cement, fly ash, cement kiln dust, or different proprietary chemical product. Bituminous stabilization is achieved by using asphalt binder, asphalt emulsion, and foam asphalt. A combination of these methods is also used for better stabilization.

### 3.1 Combining all layers—Full depth reclamation

#### *3.1.1 Using cement*

The widely used chemical type stabilization is cementitious agents such as lime, Portland cement, lime, type C fly ash, and blends materials. Stabilizing agents increase the particles' bonding and reduces the materials' plasticity. Lime stabilizing agent is typically used when the Plasticity Index of RAP is more than 10. Meanwhile, Portland cement is more often used if the Plasticity Index is less than 10.

The more stabilizing agent will increase the strength gain by the reclaimed material, but it is not necessarily better because increased amount of stabilizing agent may lead to a more rigid material. The increase in rigidity means that the reclaimed material is less prone to fatigue failure but has less performance against shrinkage cracking. However, shrinkage cracking related to cementitious materials can be alleviated by several methods such as lowering the application content but aligning with the mix design, maintaining compaction of the reclaimed mix to be below 75% saturation or below the optimum moisture content and control the rate of drying of the reclaimed mix. To better evaluate the amount of stabilizing agent, the physical properties of reclaimed mix with chemical stabilizing agent are related to the overall pavement design.

### *3.1.2 Using emulsion*

Recent advances in chemical technology bring bituminous stabilizing agents to gain more popularity. Bituminous stabilizing agents are used in the form of asphalt emulsion. Asphalt emulsion as a stabilizing agent enables a mix between asphalt bitumen and cold reclaimed material. The compatibility in the chemistry aspect of asphalt emulsion and reclaimed asphalt material is important to form a stable bitumen-bounded reclaimed mix.

Better fatigue properties, less shrinkage cracking, and faster traffic opening were found in reclaimed mix with bitumen stabilizing agents compared with the reclaimed mix treated with cementitious stabilizing agents. The bituminous stabilized material behaves as a granular material with interparticle friction and as a viscoelastic material capable of withstanding repeated tensile stress. Bituminous stabilized mixes are susceptible to moisture effect due to high void ratio. The addition of Portland cement, lime, and fly ash combined with bituminous stabilizing agents can reduce moisture sensitivity. To significantly increase the strength or resistance to moisture, 1–3 percent addition by weight of reclaimed material is used. This amount of addition was found to not affect the fatigue properties and can be a catalyst to increase the gained strength of the reclaimed materials. If marginal quality is found in the reclaimed materials, it is common practice to use stabilizing agents consisting of bituminous and Portland cement or lime.

**Figure 5.**
*Pill. Core from foamed bitumen stabilized asphalt layer [44].*

### *3.1.3 Using foamed asphalt*

Foamed asphalt is a cost-effective technique for stabilizing earth material, granular material, and bituminous material. The process mixes asphalt binder and water together to create a foamed asphalt which expands the size of asphalt approximately 17 times and allows it to coat soils and aggregate. The foaming process can be applied in place or in a central plant. Schwartz et al. [48] provided a recommended default design values for resilient modulus and structural layer coefficient are 300 ksi to 400 ksi and 0.30–0.35, respectively. Maryland State Highway Agency developed and adopted a standard specification Section 513—FOAMED ASPHALT STABILIZED BASE COURSE to provide best practice of quality control, equipment, acceptance for placing foamed Asphalt Stabilized Base Course. Typically, foamed asphalt will generate slight darkening of the new mixture as shown in **Figure 5**. This is due to uncoated aggregate and less bitumen found in foamed asphalt stabilized mixture. Foamed asphalt adheres to the fine particles and acts as a mortar that binding the particles. The advantages of foamed asphalt are the mix can work in various weather conditions and the asphalt can immediately be available for traffic after compaction.

## 3.2 Removing and treating individual layers

Instead of combining bituminous and non-bituminous layers, another approach treats and replaces individual layers. An example to that is the removal/milling of asphalt layers, piling the milled asphalt nearby and using a mobile central plant (such as Wirtgen's KMA 220) to process the material in the project vicinity. The material is then treated (with foam, cement, or emulsion as described in Section 3.1 then transported and placed. The advantage of this process is to allow a more precise control of pavement layers and utilizing the conventional layering system. **Figure 6** shows an example of a multi-step process where the top layer is milled out and processed using a central plant, the base is in place recycled then the plant processed asphalt layer is placed on top of the recycled base.

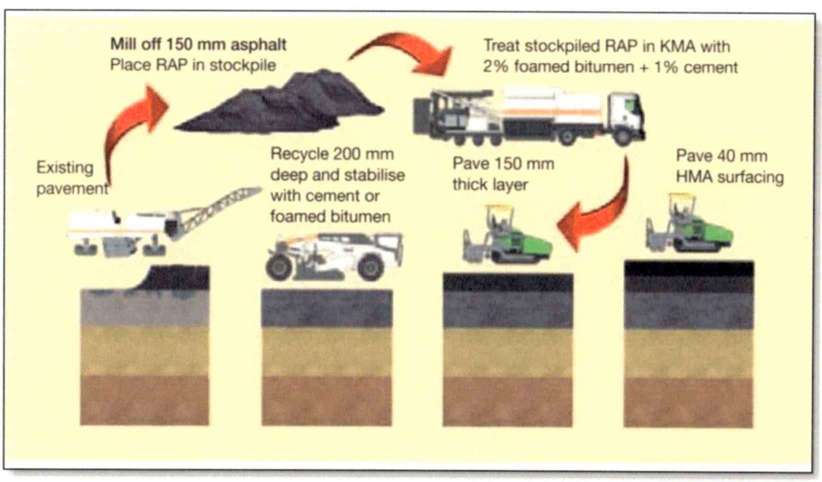


**Figure 6.**
*Example of a 2-step process to peel and recycle individual layers [46].*

Virginia Department of Transportation pioneered the use of pavement recycling techniques referenced above in high class roads, Interstate I-81, and used three different techniques: full-depth reclamation, cold centralplant recycling, and cold in-place recycling in the project. In order to accomplish the work, VDOT used a unique lane-closure schedule. An approximately eight months construction contract with valuation of $7.64 million was awarded in December 2010. VDOT published several reports and papers and provided several subsequent case studies for the nation on the use of best recycling techniques.

## 4. Conclusions

Pavement recycling is a clear way to achieve sustainability. This chapter provides a brief introduction and outlines the state of practice with respect to asphalt pavement recycling. The reuse of various pavement layers (asphalt, asphaltic base, granular base, and subbase) is discussed. The ways of utilizing the material, from the highest to the lowest economical use, are presented. The methods of recycling individual layers or combining multiple layers are discussed.

## Acknowledgements

Authors acknowledge the help from various organizations to develop this paper. This includes Wirtgen America, Florida Department of Transportation, Maryland State Highway Agency, and the Asphalt Recycling and Reclaiming Association.

## Conflict of interest

The authors declare no conflict of interest.

## Author details

Hesham Ali[1*] and Aditia Rojali[2,3]

1 Sustainable Road Engineering, Sunrise, FL, USA

2 Florida International University, Miami, FL, USA

3 Universitas Ibnu Khaldun, Bogor, Indonesia

*Address all correspondence to: hesham123@comcast.net

## References

[1] Transportation Research Board. Application of Reclaimed Asphalt Pavement and Recycled Asphalt Shingles in Hot-Mix Asphalt: National and International Perspectives on Current Practices. Transp Res Circ; 2014; E-C 188(October). p. 78. Available from: http://onlinepubs.trb.org/onlinepubs/circulars/ec188.pdf

[2] Dubravac M, Softić E, Talić Z. Analysis of Economic Feasibility and Usefulness of Asphalt Mixtures of Recycled Asphalt in Relation to the New Ones. Vol. 28. Lect Notes Networks Syst; 2018. pp. 513-523

[3] Zaumanis M, Mallick RB. Review of very high-content reclaimed asphalt use in plant-produced pavements: State of the art. International Journal of Pavement Engineering. 2015;**16**(1):39-55. DOI: 10.1080/10298436.2014.893331

[4] Williams BA, Willis JR, Shacat J. Annual Asphalt Pavement Industry Survey on Recycled Materials and Warm-Mix Asphalt Usage: 2019. Natl Asph Pavement Assoc.; 2020. p. 48

[5] EAPA. The Use of Secondary Materials, By-Products and Waste in Asphalt Mixtures. Brussels; 2020. Available from: https://eapa.org/eapa-position-papers/

[6] Lin J, Hong J, Xiao Y. Dynamic characteristics of 100% cold recycled asphalt mixture using asphalt emulsion and cement. Journal of Cleaner Production. 2017;**156**:337-344. DOI: 10.1016/j.jclepro.2017.04.065

[7] Perez F, Rodriguez M, De Visscher J, Vanelstraete A, De Bock L. Design and performance of hot mix asphalts with high percentages of reclaimed asphalt: approach followed in the paramix project. In: Proc 3rd Euroasphalt Eurobitume Congr. 2004;(CONF)

[8] Colbert B, You Z. The determination of mechanical performance of laboratory produced hot mix asphalt mixtures using controlled RAP and virgin aggregate size fractions. Construction and Building Materials. 2012;**26**(1):655-662. DOI: 10.1016/j.conbuildmat.2011.06.068

[9] LGDP S, AMDC B, Capitão SD. Assessment of the use of hot-mix recycled asphalt concrete in plant. Journal of Transportation Engineering. 2010;**136**(12):1159-1164

[10] Silva HMRD, Oliveira JRM, Jesus CMG. Are totally recycled hot mix asphalts a sustainable alternative for road paving? Resources, Conservation and Recycling. 2012;**60**:38-48. DOI: 10.1016/j.resconrec.2011.11.013

[11] Al-Qadi IL, Elseifi M, Carpenter SH. Reclaimed asphalt pavement - A literature review. In: Report No. FHWA-ICT-07-001. Fed Highw Adm. Vol. 07. 2007. pp. 1-25

[12] Tabaković A, Gibney A, McNally C, Gilchrist MD. Influence of recycled asphalt pavement on fatigue performance of asphalt concrete base courses. Journal of Materials in Civil Engineering. 2010;**22**(6):643-650

[13] McDaniel RS, Shah A, Huber GA, Copeland A. Effects of reclaimed asphalt pavement content and virgin binder grade on properties of plant produced mixtures. Road Materials and Pavement Design. 2012;**13**(Suppl. 1):161-182

[14] Yang R, Kang S, Ozer H, Al-Qadi IL. Environmental and economic analyses of

recycled asphalt concrete mixtures based on material production and potential performance. Resources, Conservation & Recycling. 2015;**104**:141-151. DOI: 10.1016/j.resconrec.2015.08.014

[15] Mogawer W, Bennert T, Daniel JS, Bonaquist R, Austerman A, Booshehrian A. Performance characteristics of plant produced high RAP mixtures. Road Materials and Pavement Design. 2012;**13**(Suppl. 1):183-208

[16] Izaks R, Haritonovs V, Klasa I, Zaumanis M. Hot mix Asphalt with high RAP content. Procedia Engineering. 2015;**114**:676-684. DOI: 10.1016/j.proeng.2015.08.009

[17] Stroup-Gardiner M. Use of Reclaimed Asphalt Pavement and Recycled Asphalt Shingles in Asphalt Mixtures. A Synthesi. Transportation Research Board, editor. NCHRP SYNTHESIS 495. Washington D.C.: The National Academies Press; 2016. pp. 1-122

[18] Copeland A. Reclaimed Asphalt Pavement in Asphalt Mixtures: State of the Practice. Report No. FHWA-HRT-11-021. Virginia, USA; 2011 Available from: https://www.fhwa.dot.gov/publications/research/infrastructure/pavements/11021/11021.pdf

[19] El-Badawy SM, Gabr AR, El-Hakim RTA. Recycled materials and by-products for pavement construction. In: Handb Ecomater. Vol. 3. 2019. pp. 2177-2198

[20] Chesner WH, Collins RJMM. User Guidelines for Waste and by-Product Materials in Pavement Construction. FHWA-RD-97-148. Washington DC: FHWA; 1998. Available from: https://www.fhwa.dot.gov/publications/research/infrastructure/structures/97148/cfa53.cfm

[21] Muñoz Perez SP, Onofre Maicelo PAA. Use of recycled asphalt as an aggregate for asphalt mixtures: literary review. Innovative Infrastructure Solutions. 2021;**6**(3):1-11. DOI: 10.1007/s41062-021-00516-x

[22] Naulkha U, Bathla A. A review study on use of reclaimed Asphalt Pavement (RAP) materials in flexible pavements. International Journal of Development Research. 2018;**6**(1):1

[23] Asphalt Recycling and Reclaiming Association (ARRA). Basic Asphalt Recycling Manual. 1st ed. Washington D.C.: FHWA; 2001

[24] Pradyumna TA, Jain PK. Use of RAP stabilized by hot mix recycling agents in bituminous road construction. Transportation Research Procedia. 2016;**17**(December 2014):460-467. DOI: 10.1016/j.trpro.2016.11.090

[25] Tarsi G, Tataranni P, Sangiorgi C. The challenges of using reclaimed asphalt pavement for new asphalt mixtures: A review. Materials (Basel). 2020;**13**(18)

[26] Tran B, Hassan R. Performance of hot-mix asphalt containing recycled asphalt pavement. Transportation Research Record. 2011;**2205**:121-129

[27] Mousa E, Azam A, El-Shabrawy M, El-Badawy SM. Laboratory characterization of reclaimed asphalt pavement for road construction in Egypt. Canadian Journal of Civil Engineering. 2017;**44**(6):417-425

[28] Collins RJ, Ciesielski SK. Synthesis of Highway Practice 199: Recycling and Use of Waste Materials and By-Products in Highway Construction. 1994

[29] Saeed A. Performance-Related Tests of Recycled Aggregates for Use in Unbound Pavement Layers. NCHRP

REPORT 598. Washington, D.C.; 2008. Available from: https://www.trb.org/Publications/Blurbs/159851.aspx

[30] Hoppe JE, Stephen D, Fitch GM, Shetty S. Feasibility of Reclaimed Asphalt Pavement (RAP) Use As Road Base and Subbase Material. Charlottesville, Virginia: A Rep Virginia Cent Transp Innov Res Univ Virginia; 2015. pp. 1-42

[31] Zaumanis M, Cavalli MC, Poulikakos LD. How not to design 100% recycled asphalt mixture using performance-based tests. Road Materials and Pavement Design. 2020;**21**(6):1634-1646. DOI: 10.1080/14680629.2018.1561381

[32] Lee J, Erik Denneman YC. Maximising the Re-use of Reclaimed Asphalt Pavement Outcomes of Year Two : RAP Mix Design. Austroads; 2015 Available from: www.austroads.com.au

[33] Dong Q, Huang B. Laboratory evaluation on resilient modulus and rate dependencies of RAP used as unbound base material. Journal of Materials in Civil Engineering. 2014;**26**(2):379-383

[34] Bennert T, Papp J, Maher A, Gucunski N. Utilization of construction and demolition debris under traffic-type loading in base and subbase applications. Transportation Research Record. 2000;**2**(1714):33-39

[35] Locander R. Analysis of Using Reclaimed Asphalt Pavement (RAP) as a Base Course Material. Denver, USA; 2009. Available from: http://www.coloradodot.info/programs/research/pdfs/2009/rapbase.pdf

[36] Taha R, Ali G, Basma A, Al-Turk O. Evaluation of reclaimed asphalt pavement aggregate in road bases and subbases. Transportation Research Record. 1998;**1652**:264-269

[37] Cosentino PJ, Kalajian EH, Bleakley AM, Diouf BS, Misilo AJ, Petersen AJ, et al. Improving the Properties of Reclaimed Asphalt Pavement for Roadway Base Applications. 2012;**6975**(321):605. Available from: http://hawaiiasphalt.org/wp/wp-content/uploads/FDOT_BDK81_977-02_rpt-RAP-in-Base-Course.pdf

[38] Bleakley A, Cosentino P. Improving properties of reclaimed asphalt pavement for roadway base applications through blending and chemical stabilization. Transportation Research Record. 2013;**2335**:20-28

[39] McGarrah EJ. Evaluation of Current Practices of Reclaimed Asphalt Pavement/Virgin Aggregate as Base Course Material. Univ Washington, Seattle, 2007. 2007;(December):41.

[40] Soleimanbeigi A, Edil TB. Compressibility of recycled materials for use as highway embankment fill. Journal of Geotechnical and Geoenvironmental Engineering. 2015;**141**(5):04015011

[41] Sayed SM, Pulsifer JM, Jackson NM. UNRAP: Are we ready for it? Journal of Materials in Civil Engineering. 2011;**23**(2):188-196

[42] Saeed A, Hammons MI. Minimum standards for using recycled materials in unbound highway pavement layers. In: Airf Highw Pavements Effic Pavements Support Transp Futur - Proc 2008 Airf Highw Pavements Conf. Vol. 329. 2008. pp. 453-464

[43] Ahmed I, Lovell CW. Use of waste materials in highway construction: State of the practice and evaluation of the selected waste products. Transportation Research Record. 1992;**1345**:1-9. Available from: https://onlinepubs.trb.org/Onlinepubs/trr/1992/1345/1345-001.pdf

[44] Updyke BE, Ruh D. Abundance of RAP Spurs new uses in Preservation treatments low-dust application. Pavement Preservation Journal. 2016. Available from: https://www.pavementpreservation.org/wp-content/uploads/2016/12/FPPQ_FPPQ0416_Page_25.pdf

[45] Tarefder RA, Ahmad M. Cost-effectiveness analysis of chip seal with and without millings. International Journal of Pavement Engineering. 2018;**19**(10):893-900. DOI: 10.1080/10298436.2016.1219599

[46] Wirtgen GmbH. In: Group W, editor. Wirtgen cold recycling manual. 2nd ed. Windhagen, Germany: Wirtgen GmbH; 2004. 251 p

[47] Jahren CT, Cawley B, Ellsworth B, Bergeson KL. Review of Cold In-Place Asphalt Recycling in Iowa. Transp Conf Proc. 1998;(Figure 1). pp. 259-263

[48] Schwartz CW, Khosravifar S. Design and Evaluation of Foamed Asphalt Base Materials. Baltimore, Maryland, USA; 2013. Available from: https://roads.maryland.gov/OPR_R

Section 2

# Circular Economy and Waste Challenges

Chapter 7

# Circular Economy: An Antidote to Municipal Solid Waste Challenges in Zambia

*Kachikoti Banda, Erastus M. Mwanaumo and Bupe Getrude Mwanza*

## Abstract

Zambia is one of the fastest developing countries in Africa. It is land linked and has one of the most urbanizing cities, the capital, Lusaka. The country is now grappling with serious challenges of managing municipal solid waste that is generated from its growing population and increased economic activity. Circular economy ensures that all the negativities of linear economy are reduced or prevented by ensuring reduced generation of waste at source, reuse of the generated waste and if these cannot be implemented, recycling of the generated waste follows. This results into environmental benefits such as clean and safe air and water. Land degradation or pollution is prevented. Therefore, there is need to implement circular economy as an antidote to the current municipal solid waste challenges. Municipal solid waste management is a critical public good that provides a barometer for the effectiveness of any governance system around the world. Successive governments should embed the waste management issue in all the policies developed for to ensure sustainability. In today's world of material scarcity and a call to action toward climate change action, it cannot be over emphasized that circular economy is the antidote to municipal solid waste challenges Zambia is facing.

**Keywords:** circular economy, municipal, solid waste, antidote, linear economy

## 1. Introduction

Rising populations and increased economic activities across the world have given rise to the generation of municipal solid waste. While developed countries have made tremendous slides to cope with this increased generation, developing countries especially in southern Africa are still grappling with the challenge. Developing countries are making efforts to improve the well-being of their people, grow the economy and ensure the much-needed development is delivered especially when it comes to political, promises. While they focus is on the former, municipal solid waste generation rates continue to rise and are not met with sustainable municipal solid waste management systems. This paper focuses on municipal solid waste management in Zambia, a southern African country and how circular economy could be used as an antidote or solution to many challenges the country is facing.

Zambia is a vast African country and one of the fastest developing countries in Africa and has one of the most urbanizing cities, the capital, Lusaka. National records indicate that the country covers 752,614 square kilometers and the World Bank projects the 2021 population to be at 18.9 million. According to www.cia.gov, the rate of urbanization in Zambia is estimated at 4.2% for the period 2020 to 2025 with the urban population accounting for 46% as at the year 2022. The country applies democratic governance system with its 2016 Republican Constitution providing exclusive functions for local authorities among which is municipal solid waste management (Article 147). Rising populations and the rural urban drift due to improved economy have given rise to the daunting challenge in the management of municipal solid waste in the country. Once known as the Garden City of Africa, the country is now grappling with serious challenges of managing municipal solid waste that is generated from its growing population and increased economic activity. This is exacerbated by collapsed waste management systems, lack of financing and bad behavioral attitudes. Successive governments have tried several efforts to ensure that municipal solid waste is effectively managed to ultimately protect public health and the environment [1, 2]. While this is just an example of the challenge in the capital city, the situation is similar across the country especially cosmopolitan towns along the line of rail. Lack of financing, poor behavior attitudes couple with lack of equipment and uncontrolled unplanned settlement has led to a serious challenge to Local Authorities. The situation poses a threat to public health, environment, socio-economic and to a larger extent political sector. Further, contributions from Balasubramanian [3], indicate that social, economic and health issues are some of the effects of uncollected waste. While effects of poor waste management are innumerable as they affect all sectors of human, environmental and socio-economic development, the use of linear economy focuses only on municipal solid waste systems that recognize generation, collection, transportation and disposal only. Such linear economy as revealed by Glaser et al. [4], has been in existence since the industrial revolution and has achieved economic growth. Linear economy systems have several negative impacts on the public health and the environment as they are primarily focused on economic growth in their non-holistic approach.

Sustainable municipal solid waste management is a potential tool for socio-economic development despite its primary focus on public health and environment. As the world and Zambia as member of the global village propagates then achievement of sustainable development goals, the implementation of circular economy is an urgent matter. The call for circular economy across the world is now louder and more apparent than before. This would not only ensure a clean and safe environment but also, create the much-needed jobs with more than 60% population being the youths. As opposed to linear economy, circular economy is more productive and healthier with raw materials are maintained in the production cycle and recycled [5].

## 2. Municipal solid waste management in Zambia

Zambia attained independence in 1964 on October 24th. During that period and a few years post-independence, municipal solid waste ably managed by the colonial government and new independent Zambia. This was made possible through the public health act and other related legislation. As the country embarked on liberalizing the economy, the Environmental Council of Zambia was created to attend to environmental challenges in the country especially those coming forum the mining sector which is the main revenue base for the country. This led to formulation of the waste management regulations. a few years before that, the national conservation strategy of 1985 was developed to provide policy guidance on the use of natural resources and environmental preservation. These two categories of documents formulated the basis of legal and policy framework that government municipal solid waste management in Zambia.

As the country progress on its economic reforms, the economy was liberalized leading the country into a free market economy. Increased population growth, a booming economy coupled with the rural urban drift along the cosmopolitan towns of the line of rail led to increased population growth. It is worth noting that the colonial government and the subsequent Zambian governments after independence used a tariff bundling systems for collection of waste management fees. This system collapsed on the mid-1990s when economic reforms took place, the resultant effect was piles of uncollected waste lying in the stress because municipalities could not effectively collect waste mainly because of failure to maintain the fleet and also low subscription rates form the citizens. The situation was exacerbated with low research and funding for the waste management sector leading to outbreak of diseases such as cholera especially in major cities like Lusaka the capital.

According to the constitution of Zambia, municipal solid waste management is an exclusive function of local authorities. Therefore, as the responsibility grew, the Zambia Environmental Management Agency (ZEMA), an institution created with the enactment of the environmental management act of 2011 was regulating local authorities in the management of municipal solid waste. However, in December of 2018, the government of Zambia enacted the first ever stand-alone act of parliament, the Solid Waste Regulation and Management Act of 2018. The act recognizes waste as resources and provides for formulation of utility companies to by local authorities to manage waste on their behalf. Further, the country does not have a new waste management strategic plan and is currently using the nation solid waste management strategy of 2004. Apparently, only the Lusaka City Council from the capital Lusaka has developed a waste management plan. Local authorities in the country are visibly struggling to manage municipal solid waste. Some of the reasons attributed to the glaring problems are; lack of revenue, low or no investments or incentive sin the sector by government, extremely low levels of awareness among the citizens leading to illegal dumping, burning and burying of waste etc. While the law is in a place and few convictions have been secured especially in Lusaka, many towns and cities across the country are doing very little to combat this environmental and public health threat. What is more threatening and appalling, not all the 116 municipalities in the country have engineered landfills for safe disposal of their waste. This means the whole country is using crude dumping clearly polluting the atmosphere, land and water, both surface and underground, the eminent public health threat the immediate communities living around these dumpsites cannot be over emphasized. Further, very little landfill diversion strategies are areas being done to enhance recycling and waste to energy process in a bid to promote circular economy. From the highlighted enormous challenges, it is time the country embraced circular economy for an assured sustainable future. This calls for enhanced recycling of municipal solid waste to ensure a sustained supply of material to the industry for production. The model would reduce the stress on landfills and eventually lead to manageable levels of costs for final disposal sites.

## 3. Current generation rates, streams and sources

Municipal solid waste management in Zambia is a constitutional mandate for Local Authorities. The Ministry of Local Government and Rural Development and other agencies like the Zambia Environmental Management Agency provide an oversight role to the management of solid waste in the country. The implementation of circular economy models are based on the fact that waste generated from any activity, is used as raw material for the production of other items thereby avowing the use of virgin materials. As discussed earlier, the implementation of the 3R systems is the blueprint of this model. However, it is important to know generation rates to determine the quantities of raw materials for production. For example, waste to energy systems are one the innovative technologies that use waste as a raw material for the generation of het and energy. In world affected by climate change, the scarcity of water for hydroelectricity generation is real. Further, generation of heat and electricity from fossil fuels is a serious threat to climate change. Therefore, use of municipal solid waste for this purpose is the alternative. Zambia currently has had nor empirical survey to determine generation rates, waste streams and sources which produces what kind of waste. Expert opinion, experience and physical characterization of waste informs that for the city of Lusaka in Zambia, 60% of waste produced is comprised of plastic and paper. This could be similar for other cosmopolitan towns and cities but notably different for rural districts, which have agricultural activities and lack of manufacturing industries, low commercial activities and low production more than these towns. This implies that there is huge potential for recycling. Metal is heavily recycled in the currently because of a lucrative market for scrap metal. Glass, garments and organic waste are rarely recycled. It is therefore important that before any circular economy model is implemented, these studies be conducted to inform decision and devise strategies.

## 4. Circular economy and municipal solid waste management

According to Zhang et al. [6], Circular Economy is an economic development model consisting of resources-products-renewable resources and repeated circulation of materials based on the principles of reduce, reuse and recycle. Natural resource depletion and increased waste generation have contributed to the emergence of the concept of a "circular economy" as a new paradigm opposed to the standard "linear economy" [7]. While there are other schools of thoughts that have more than 3Rs, the scope of this paper is just the 3Rs. This means that the waste generation is reduced at source from every day economic, domestic and commercial activities. Reduction of generation of waste is a conscious-based activity at individual level and also system-based at industrial or commercial level. This means that individuals, residents, or the citizenry is well aware that they have to reduce the generation of more waste by employing technics at individual level that could help in reducing waste generation. For example, an individual or a family could use are reusable shopping bag other than purchasing more plastic bags as they con duct their shopping. Another example at commercial or industrial levels is implementation of systems that reduce more generation of waste are employed. For example, when an organization is conducting a workshop, the institutional policy to reduce the generation of waste from purchase of individual water drinking bottles could clearly be spelt such that all workshop participants could come with their own reusable water bottles and drink from the common water dispenser or reusable cups could be used for the participants. Clearly, these technics could help reduce the generation of waste that affecting the first R for reduce. The unavoidable waste that is generated could then be disposed of in separate bins according to waste streams or indeed taken to sorting centers for separation at communal sorting stations. This leads us to the second R. However, the effectiveness of this is mainly based on the mindset of the population [8] and effectiveness of the implementation of institutional or work policies aimed at reducing the generation of municipal solid waste.

When waste is generated, there are certain streams or types of waste that could be reused without any physical or chemical change in this case recycling. The reusing of such waste is based on the primary status of its physical, chemical or biological properties. For example, waste from vegetable cuttings or food premises could be used as they are for manure or compost in the backyard gardens for growing of vegetables, this would contribute to the reduction of expenditure on fertilizers, especially if it is conducted at large scale. The world has seen the rise of proponents or enthusiasts of organic framing leading to some becoming vegetarians to save the animals. The growing of such organic food is based on the reuse and recycling of organic waste or waste that can decompose and provide that much needed nutrients to plants or animals. Reduce and Reuse of waste helps in diverting the waste from final disposal sites and thus contribute to the lifespan of the landfill. This is a closed loop systems. Plastics are the other stream of waste that can be reused in its form for several purposes. This leads us to the final R; Recycling. Recycling is the manufacture or production of goods or items by use of material that has performed a primary function and not virgin material. This implies that the raw material used in the production line is waste disposed off in another activity. Recycling of waste could save millions of money needed to purchase, transport and use of virgin material in production systems. Environmental conservation and protection is at the core of the implementation of this R because of the potential of prevention of pollution, conserves the earth and ensuring a safer and cleaner environment. Recycling is thus a game changer in industrial production, job creation, public health and environmental protection. Circular economy if well implemented could be an antidote the currently challenges the country is facing in managing its waste in most of its cosmopolitan towns and cities. This is affirmed by the work of Allevi et al. [8] who propose a waste management model in a circular economy framework.

## 5. Environmental and socio-economic value of circular economy

While the implementation of circular economy as opposed to linear economy is being encouraged and considered, it is worth noting that the circular economy has numerable environmental and socio-economic value to the implementing agent, institution, city or indeed country. Linear economy involves the use of raw materials to make new products and the waste generated is disposed off. This implies that there is no diversion from the final disposal systems, for recycling let alone, reduction and ruse. This cause pollution landfills, over consumption of natural resources thereby depleting the much-needed resources for future development, clearly going against the concept of sustainable development. Continued hauling of waste from generation to disposal sites and management of disposal sites consume a lot of fuel thereby affecting budgets of municipalities across the country. The threat of climate change, pollution and contamination of air land water cannot be overemphasized. However, circular economy ensures that all the negativities of linear economy are reduced or prevented by ensuring reduced generation of waste are source, reuse of the generated waste and if these tow cannot be implemented, recycling of the generated waste. This yields innumerable environmental benefits among which are clean and safe air and water. Land degradation or pollution is prevented.

On the other hand, agents, institutions or industries and indeed municipalities are bale to record significant savings on their financials in circular economy is implemented. For example, the cost of running landfills is significantly reduced if waste is diverted to recycling plants as opposed to just being disposed of at landfills. This is true also at domestic or industrial level as evidenced by significant savings on the cost of virgin raw material needed for production. Individuals of agents can save at personal or domestic level if the 3Rs are implemented as circular economy principles.

## 6. Circular economy and vision 2030

Zambia seeks to grow its economy to achieve a middle-income prosperous nation by the year 2030. This aspiration has led to the development and implementation of the Zambia Vision 2030 policy that would spur the country to the desired status by the year 2030. Implementation of circular economy through the 3R system can greatly contribute to the achievement of these visions by the year 2030. Concerning waste management, the Vision 2030 clearly seeks to *achieve and sustain efficiency and effectiveness in the delivery of Public Services; and also attract and retain quality technical, professional and managerial staff in the Public Service*. This implies that as an antidote to the current municipal solid waste challenges the country is facing, there is need to ensure that there is effectiveness and efficiency in the implementation of the 3R system to realize the accrued benefits. Further, these results cannot be achieved if the country does not produce through its universities and colleges, technical and professional staff to undertake this important program. Often times, countries or municipalities especially in developing countries do not have experts in waste management leading to the sector being managed or handled by all sorts of people thereby causing chaos and disjointed activities whose results and impacts cannot not be measured. Therefore, circular economy, through the implementation of the 3R system in Zambia can greatly contribute to the achievement of the Government's vision for a middle-income prosperous nation by the year 2030.

## 7. Strategies for Zambia to achieve a circular economy

Zambia is party to international organizations and treaties among the notable ones being the sustainable development goals. It is therefore important that the country migrate to a fully-fledged circular economy in order to contribute to the achievement of the vision 2030 and actualize the sustainable development goals. Therefore, the following are proposed strategies to be implemented in an effort to implement circular economy as an antidote to the currently municipal solid waste challenges;

There is need for government to immediately provide incentives the waste management sector so that it becomes economically viable to reduce, reuse and recycle including waste to energy projects. Currently, there is no incentive in the waste management sector making infrastructure, equipment and collection systems economically and unattractive. The second strategy is to ensure that domestic and institutional policies are developed and enforced aimed at implementing the 3R system. Thirdly, there is need to streamline flow of all waste materials to industries and track the recycling rates for purposes of measuring progress and impacts. The last strategy is to ensure application of technology in all process so as to crate the closes loop systems for monitoring and evaluation from generation, collection, transportation, intermediate treatment, and final disposal i.e. Waste to energy plant.

## 8. Conclusion

Municipal solid waste management is a critical public good that provides a barometer for the effectiveness of any governance system around the world. Waste management is always a political issue as much as it is an environmental and public health issue. It is therefore important that successive governments should embed the waste management issue in all the policies developed for development. Further, the education and financial system should supplement enforcement and operational solutions in the sector. In today's world of material scarcity and a call to action towards climate change, it cannot be over emphasized that circular economy is the antidote to municipal solid waste challenges Zambia is facing. Environmental, social and governance factors need to be critically considered in devising waste management systems in order to combat climate change caused by municipal solid waste management [9].

## Acknowledgements

I wish to acknowledge the valuable guidance from my coach and many other unsung heroes in this work.

## Conflict of interest

The author declares no conflict of interest.

## Notes/thanks/other declarations

I am grateful to my family for their sacrifice of time during the late nights and weekends of writing. Special gratitude to my son Kachikoti whose time I always sacrificed!

## Author details

Kachikoti Banda[1*], Erastus M. Mwanaumo[1] and Bupe Getrude Mwanza[2]

1 Department of Civil and Environmental Engineering, University of Zambia, Lusaka, Zambia

2 Graduate School of Business, University of Zambia, Lusaka, Zambia

*Address all correspondence to: billkachikoti@gmail.com

## References

[1] Available from: www.lusakatimes.com/2013/07/12/lusaka-has-lost-garden-city-concept-mayor-chisenga/

[2] Zambia: 60% of garbage in Lusaka is uncollected (lusakatimes.com)

[3] Balasubramanian M. Economics of solid waste management: A review. In: Saleh HM, editor. Strategies of Sustainable Solid Waste Management. London: IntechOpen; 2020. DOI: 10.5772/intechopen.95343

[4] Glaser JA, Sahle-Demessie E, Richardson TL. Are reliable and emerging technologies available for plastic recycling in a circular economy? In: Achilias DS, editor. Waste Material Recycling in the Circular Economy—Challenges and Developments. London: IntechOpen; 2022. DOI: 10.5772/intechopen.101350

[5] Musarat MA et al. Circular economy—Recent advances in sustainable construction waste management. In: Zhang T, editor. The Circular Economy—Recent Advances in Sustainable Waste Management. London: IntechOpen; 2022. DOI: 10.5772/intechopen.105050

[6] Zhang Y, Niu Y, Zhang T. Introductory chapter: The overview of recent advances of sustainable waste management. In: Zhang T, editor. The Circular Economy—Recent Advances in Sustainable Waste Management. London: IntechOpen; 2022. DOI: 10.5772/intechopen.105574

[7] Vargas-Terranova C-A, Rodrigo-Ilarri J, Rodrigo-Clavero M-E, Parra-Saad A. Implementing circular economy techniques for the optimal management of recyclable solid waste using the M-GRCT decision support model. Applied Sciences. 2022;**12**:8072. DOI: 10.3390/app12168072

[8] Allevi E, Gnudi A, Konnov IV, Oggioni G. Municipal solid waste management in circular economy: A sequential optimization model. Energy Economics. 2021;**100**. Available from: https://www.sciencedirect.com/science/article/pii/S014098832100284X [Accessed: November 30, 2022]

[9] Jinga P. The increasing importance of environmental, social and governance (ESG) investing in combating climate change. In: Tiefenbacher JP, editor. Environmental Management—Pollution, Habitat, Ecology, and Sustainability. London: IntechOpen; 2021. DOI: 10.5772/intechopen.98345

Chapter 8

# Design for Recycling of E-Textiles: Current Issues of Recycling of Products Combining Electronics and Textiles and Implications for a Circular Design Approach

*Elisabeth Eppinger, Alina Slomkowski, Tanita Behrendt, Sigrid Rotzler and Max Marwede*

## Abstract

Circular economy principles and eco-design guidelines such as design for recycling gain increasing importance to improve recyclability of products. The market of textiles with electronic components—so-called electronic textiles (e-textiles)–grows quickly entailing an increase in waste due to obsolete and defect products. This chapter presents insights into the current state of e-textile recycling in Europe. As electronic recycling differs from textile recycling, a survey of sorting and recycling businesses in Europe was conducted to obtain insights into the current and future handling of e-textiles. The survey results reveal that e-textiles have so far played a minor role for sorting and recycling companies, but about one-third of the businesses already experienced issues in recycling e-textiles. While some of the respondents have already developed processing concepts, the overall occurrence of e-textiles is so low that businesses are unlikely to develop recycling solutions. However, with increasing market volume, waste will also increase and recycling requires improvement to reduce environmental impact. Based on the survey results, recommendations for improving the recyclability and recycling rate of e-textiles are proposed. This includes expanding the scope of current regulations to e-textiles to apply guidelines for integrating sustainable end-of-life solutions in the product design process, acknowledging current shortcomings of the recycling industry.

**Keywords:** circular design for e-textiles, eco-design, electronics recycling, e-textiles, textile recycling

## 1. Introduction

E-textiles experience an increasing popularity in both consumer product markets as well as the research community. The global market volume of e-textiles is expected to more than double from 2021 by 2026 [1]. Over the last decades, research and development of e-textiles focused on increasing the wearing comfort, robustness, reliability, and cleaning ability of the products as these properties are crucial for user acceptance [2]. One way to improve these properties is to miniaturize the electrical circuitry and fuse it to or combine it with the textile substrate [3]. However, this trend toward a high degree of integration of the electrical components into textiles leads to challenges in terms of reparability and recycling of e-textiles and their components [4]. Recycling of textile products due to different fiber material combinations, auxiliaries, such as buttons and zippers, and various chemical treatments is already difficult and hardly practiced. Also recycling of electronics because of different, strongly connected materials is challenging. Due to their hybrid nature, recycling and recovery of reusable resources from e-textiles are even more challenging than for pure textile or electronic products. Particularly, because waste collection and recycling businesses are either specialized on textiles or on electronic products.

Hence, this chapter aims to shed light onto the current processing of e-textiles in sorting and recycling companies within Europe. It provides insights into product features and conditions that must be met to ensure the recycling of e-textiles. Furthermore, it provides recommendations on conditions that have to be fulfilled to develop circular product life cycles of e-textiles. The insights into the current state of recycling and waste management for e-textiles within Europe, including challenges and possible solutions, are based on a survey among sorting and recycling companies in the textile and electronics sector conducted in the year 2021.

Circular economy principles include reuse, repair, and recycling [5]. In order to improve the reuse, reparability, and recycling of e-textiles, circular design principles should already be incorporated in the design stage. But how can we achieve this effectively for e-textiles? Based on the results of the survey, we propose that it needs to be governed top-down. A bottom-up approach from the industry is not likely to happen due to the small quantities of e-textiles.

The European Union (EU) committed its member states to sustainability transitions of manufacturing industries, among other strategies by moving toward a circular economy [6]. Other regions and states, such as China, India, Japan, and the United States of America, strengthen their commitments to circular economy and sustainable manufacturing [7–9]. Consequently, the results of this study are relevant to other territories, which are committed to sustainability transitions of manufacturing industries. As the United Nations Sustainable Development Goals (UNSDG) gain importance around the world for transforming economic activities to sustainability, the results of this study presented in this chapter contribute to UNSDG 9—Industry, Innovation and Infrastructure, and UNSDG 12—responsible consumption and production.

The chapter is structured as follows: Section 2 provides a short introduction into e-textiles and their current market development. Section 3 explicates the survey method and data that was conducted to obtain insights into the current and future handling of e-textiles by sorting and recycling companies. Section 4 presents the results of the survey, including a discussion. Section 5 summarizes the main conclusions and proposed ways to improve recyclability of e-textiles.

## 2. Current trends and challenges in product design, markets, and recycling of E-textiles

E-textiles are combinations of electronics and textiles. They consist of electronic components such as circuits, sensors, and lights for achieving functionalities of garments and textile products. Application examples include conductive components for sensing and actuating, communicating, and microprocessing information such as acoustic and motion signals [10]. The products range from monitoring health status of patients, tracking body functions, speed and routes for personal feedback in professional sports products, at leisure sport and fitness exercises, safety performance such as light signals, to acoustic combinations connectable to mobile devices for leisure [1, 3, 10, 11]. Against the backdrop of the growing market of e-textiles with a global turnover expectation of about US$ 1.3 billion by 2032 [12], the e-textile waste will increase accordingly.

Current research reveals that the shortcomings of durable integration of electronics and a lack of standards to analyze the performance are still major reasons for market failure [13, 14]. Test standards are still under development [1]. A current trend to improve longevity of e-textiles is through miniaturizing electrical circuitry and combining it with the textile substrate. The electronic components need to be fixed onto or inside the textile structures, which is done through different processes such as gluing, welding, brazing, and soldering. These joints between rigid electronic components and flexible textile substrates are often the critical product feature that is most likely to fail. A current development trend is to integrate conductive threads through knitting, stitching, sewing, and weaving into fabrics as well as printing circuits onto textile structures, instead of usage of conventional electronic components such as cables and circuit boards mounted on hard plastics [15, 16]. These electric conductive textile substrates, also known as fiber-based devices, appear to be promising to enable comfortable and durable solutions [17], and improving the washability of e-textiles [18]. However, a high degree of integration of the electrical components leads to challenges in terms of reparability and recycling of e-textiles.

Electronic waste and textile waste have different collection and recycling systems that differ among European countries. In Germany, for example, waste collection falls under communal services, and municipalities work with a variety of private businesses and charity organizations to enable collecting, sorting, recycling, and resale. While consumers bring back defect electronic and electrical products to where they have purchased them, lately, also fashion brands take on more responsibility in actively communicating to consumers that they can dispose their used items at the producers and offer take-back systems [19]. Overall, the textile and electronic waste collection in the EU is still under development, with electronic waste collection being more advanced than textile waste collection and high percentages being disposed in household trash and end up in landfills.

Electronic recycling and textile recycling both face several issues. Both require careful sorting to guarantee efficient and high-quality recycling, and for both product groups, the recycling rate is rather low, given the complex product compositions. As for textile recycling, different fiber compositions and auxiliaries make automation in sorting and processing rather challenging. The mandatory product labels that indicate the raw material compositions are often cut out by consumers, wear out through washing, or are wrongly labeled right from the start [20]. While the overall recycling rate is rather low including packaging and other waste, the EU sets targets to increase recycling and recovery of resources. For textiles, the low cost supply of new raw materials is a serious barrier to improve and increase recycling [19]. The current business model of textile sorting and recycling businesses is based on resales of high-quality second-hand garments. With decreasing quality of used textiles and a steep increase in second-hand markets that are organized through internet platforms and enable users to directly sell their apparel, the waste recycling businesses require new sources of income to sustain [19, 21].

The EU has strict regulations on the treatment of waste in general, which is continuously adjusted and guides the waste treatment in the member countries. With the aim to transition to sustainable manufacturing and consumption, further initiatives on EU and on national level guide the treatment of textile and electronic waste that facilitate recycling. However, not all countries have implemented the stricter recycling and recovery regulations [22]. The treatment of electronic components falls under the EU directive on waste electrical and electronic equipment (WEEE Directive) [23]. Along with stricter electronic and textile waste regulation, especially for disposal of WEEE criminal activities increased, resulting in illegal landfills, which primarily occurred in low-income countries [24]. This development reveals the issues at stake, that with increasing costs for waste treatment, the institutions to enforce proper waste treatments need to be strengthened as well to counteract false disposal. Hence, concepts and technologies for waste treatment should be advanced and implemented on a global scale to be effective [25].

## 3. E-textile recycling in Europe: survey method and data

The two main components of an e-textile are the textile substrate and the electronic and electrically conductive components. Based on this, two target groups were defined for the survey: (1) electronics collectors and recycling businesses, and (2) textile collectors and recycling businesses. Applying a purposive sampling strategy, using industry websites, 506 businesses from electronics recycling, 179 from textile recycling, and 81 with cross-sector expertise were identified and requested to participate per email, in English and in German language. The processing time of the survey was approximately 15 min, which was conducted online using the survey tool "LimeSurvey."

In total, the survey consisted of 46 questions for electronics and 47 questions for textiles. According to the prior experience with e-textiles, not all questions required a response with about 10 questions being optional. The questions addressed the extent to which e-textiles are recognized in sorting and recycling companies within Europe and already processed in a suitable way. The survey included a list of product features and conditions that may facilitate the recycling of e-textiles. The features and conditions were identified based on an extensive literature research and discussions with recycling experts. The option to add further characteristics and conditions enabled respondents to expand the predefined list.

To increase the response rate, the survey was carried out anonymously. The survey has a coverage bias as it was conducted online, which requires a stable internet connection, and the companies needed to have an email address, which excludes smaller sorting and recycling businesses in rural regions. A response bias may also occur due to a desired external presentation of the company or in the course of internal confidentiality or security agreements, despite the fact that the survey has been anonymized. Politeness or the desire to complete the survey despite a lack of expertise can also contribute to a bias in the results [26]. The non-response bias includes decisions not to participate or stop participation. Direct declines gave usually the lack of time due to the persistence of the Covid-19 pandemic as their reason. It is conceivable that smaller businesses in particular did not participate in the survey due to a lack of time and personnel.

The data were collected during four weeks in June 2021. The response rate of complete datasets was 6.13%. With regard to sector classification, a bit more than half of the respondents (57%) belong to the sorting and recycling sector in the field of used textiles, the others operate sorting and recycling of e-waste. Within the EU, most processing facilities of the companies that participated in the survey are located in Germany, followed by Lithuania, Poland, Spain, and Hungary. About 6.7% of the companies operate sorting and recycling facilities outside the EU, such as in Macedonia, Switzerland, and the United Arab Emirates. Outsourcing outside of the EU occurs only for textiles and not for electronics. In terms of how well the respondents represent the recycling industry in the EU, it is important to note that a large cluster of textile sorting and recycling businesses exists in East European countries [27]. However, the purchasing ability in Western European countries is higher; accordingly, a higher amount of e-textile waste can be assumed in the dominant region of the respondents. Also, the textile waste collection and recycling facilities are very structured in Germany, as compared with other countries [28]. Hence, with the majority of respondents from Germany, the results could be interpreted in terms of perspectives from advanced sorting and recycling regions.

The respondents that operate in electronic waste claimed to sort WEEE into defined categories (37%), disassemble them into their components (22%), and process WEEE as second-hand goods or recycle it mechanically (19%). Only one business claimed to recover precious metals. However, all of the respondents forward the waste at least partially to external companies for recycling and recovery processes and for reuse as second-hand goods. None of the respondents forward the WEEE outside of the EU. This might be due to stricter regulations of the electronic waste trade as compared with used textiles.

Regarding used textiles, 32% of the respondents stated that they sort into defined categories, and about 26% reprocess the used textiles as second-hand goods. Mechanically recycling of textiles is done by 23%, and only 13% of the companies indicated that they use thermal recycling for energy recovery. However, as thermal and mechanical recycling is the most common approach, this may be outsourced to others, so that the respondents do not do it themselves. In fact, about 75% of the businesses that sort used textiles exclusively have outsourced processing operations to others, such as recycling, recovery, or reprocessing into second-hand goods.

## 4. Current state of E-textile recycling: survey results and discussion

The study is guided by the question, whether e-textiles are recognized in sorting and recycling companies within the EU, and processed in a suitable way that addresses both the electronic and the textile components. Accordingly, the study contained a question about the occurrence of e-textiles in the recycling process. The sorting and recycling companies responded that currently the quantities of e-textile waste that show up at their businesses are rather low to very low. The e-textiles originate from textile container collections (33%), from hospitals and industrial manufacturing (17%), and from municipality WEEE collection points (17%). The rest steams from unknown sources. The e-textiles had integrated, flexible printed circuit boards, and stretchable printed circuit boards, embroidered circuits, as well as integrated textile circuit boards. Whereas printed and stretchable circuit boards showed up at both subsectors (electronic recycling and textile recycling), the e-textiles with embroidered circuits and with integrated textile circuit boards only showed up in textile sorting and recycling businesses. They did not get processed except for thermal recovery, whereas the others with printed circuit boards got either forwarded to second-hand markets or separated into textile and electronic components and got processed further accordingly within electronic and within textile recycling.

Regarding existing processing concepts for e-textiles, 28.6% of the textile and electronic sorting and recycling companies stated that they already process e-textiles.

In total, 37.5% stated that they do have a processing concept. The businesses with processing concepts usually sort the waste based on the type of e-textile, high quality gets forwarded to second-hand markets and for lower quality, the components get separated. In a next step, the textile content gets mechanically separated and processed into fibers for nonwovens, for example, for cleaning rags or insulation material, or for new fabrics. Energy recovery is also quite common. The electronic sorting and recycling businesses claimed that they would recover secondary raw materials such as precious metals from the electronic parts. It is conceivable that e-textiles that have occurred in the companies to date have not been documented with details of their construction type. This makes it difficult to draw conclusions about the occurrence of different e-textile systems and their current recycling.

The respondents were not aware of any business that specialized on recycling of e-textiles. This appears plausible given the very few products that end up at recycling facilities. While 62.5% of the companies in the textile sector can reliably identify e-textiles during processing operations, 37.5% of all companies stated that they experienced difficulties during further processing of e-textiles in the past. This was among other issues due to e-textiles that remained undetected. The undetected e-textiles caused reduction of the quality in terms of purity grade of the recycling streams. Accidental shredding of these components contaminates the recycling streams. One respondent explained that there may be an additional fire hazard. The respondents confirmed that a high integration of electronics and textiles is difficult in recycling. Especially permanent bonds between the textile substrate and conductive yarns such as for e-textiles with embroidered circuits cannot be disassembled so far. Consequently, they get sorted out for thermal recovery.

The future emergence of e-textiles is estimated to be rather low and very low by the majority of the respondents (82.3%). Although a strong market increase of e-textiles is forecasted for the next decade [1], it will continue to make up only a small segment of the overall textile and electronics markets in the future. Thus, the market volumes of regular textile products and electronics will continue to exceed the market volume of e-textiles by a multiple. Given the low volume of market-ready e-textiles, there does not seem to be any urgency for recyclers to develop specialized processing methods at this stage. It is likely that processing methods specifically designed for e-textiles are not yet economically viable or cost-covering for the companies. This may result in lack of action, with the industry failing to develop efficient solutions for e-textiles. Accordingly, it should be governed top-down by policymakers, as the quantities are too low for industry to develop bottom-up solutions.

To recycle e-textiles efficiently, the companies explicated various requirements and conditions. These requirements include product design that allow easy disassembly, reliable identification of electronic components, clear waste regulations to help consumers understand how to dispose e-textile waste properly, the development of integrated factories that can process both textiles and electronic components, the documentation and evaluation of processes for reliable data, and a general improvement of the recycling processes of electronic waste and textile waste. Electronic and textile waste both operate on a very low margin, and the processes to regain high-quality resources for further products are still expensive.

In order to identify e-textiles in sorting, the product labeling regulations could include specific markers. In case of non-detection, sorting companies exporting end-of-life textiles to third countries run the risk of exporting e-textiles together with other textile waste. This may constitute an illegal export of e-waste to third countries. In the survey, we asked about feasible ways to mark e-textiles from a sorting and recycling perspective. The respondents had different views on the practicality of various markings to facilitate reliable identification of e-textiles. About 11% stated that standardized markings were not necessary for the identification of e-textiles and that a visual inspection was sufficient. About 22% found a text on the sewn-in tag in the product practicable, and about 22% considered RFID tags useful. In total, 19% voted for printed or embroidered text on the textile surface, whereas 15% found an embroidered or printed QR code on the textile surface the best solution, and 11% selected QR codes onto the sewn-in tag or color stripes. The use of chemical marker was also mentioned as an alternative solution for efficient and reliable identification of e-textiles. This can enable time-efficient detection of e-textiles in the near-infrared range, which would eliminate the need to search for a marker. Overall, the variety of answers reflect the uncertainty and need for a practicable solution to mark e-textiles.

Product marking with RFID tags adds an additional microchip and antenna-based electrical component that must be properly processed at the end of the product's life. To find out whether the textile sorting and recycling businesses have already experienced difficulties with RFID-tagged products, the companies were surveyed in this regard. About 28% reported that difficulties already occurred due to integrated RFID tags during processing operations. Specifically, difficulties arose in sorting products correctly according to their RFID tags. In addition, one business indicated that problems were suspected to occur during mechanical and chemical recycling processes. One business from the electronic sorting and recycling sector stated that there was no RFID detection in primary treatment plants and that the sensor technology in sorting plants can react to RFID tags with error messages.

The majority of respondents (about 81%) agree that special collection systems for discarded e-textiles would support proper recycling. However, the remaining 19% disagree with the statement. The issue that users dispose their waste incorrectly despite collection systems and awareness campaigns could be the reason why the respondents have rather different views on the value of specially dedicated collection system for e-textiles. The other reason may be the low quantity of e-textiles, which hardly justifies dedicated collection systems. Hence, take-back solutions by producers or disposal at WEEE collection points at municipalities are likely to be sufficient.

Waste regulations need to be combined with campaigns to inform users of e-textiles. To efficiently process end-of-life e-textiles, end users need sufficient knowledge about the proper disposal. As the quantity of e-textile waste is still very low, we asked in the study about the sufficiency of knowledge on proper disposal of textiles and electronics. The different responses from textile sorting and recycling businesses as compared with electronic sorting and recycling show that knowledge of proper textile waste disposal is lower. Regarding used textiles, only 12.5% of the respondents that operate in textile waste rated the knowledge of end users for the proper disposal as rather sufficient. About 68.75% stated that the existing knowledge of end users is rather insufficient, and the remaining 18.75% claimed that end users lack appropriate knowledge. The knowledge how to correctly dispose WEEE appears slightly better with 50% of the electronic waste treatment businesses considered the knowledge of end users to be rather sufficient. Only 33% find the knowledge rather insufficient, and again the remaining 17% rate it as insufficient. The knowledge that wrong disposals of electronic and electrical components pollute the environment and that the contained metals should be recovered is probably more widespread than the consequences of disposing used textiles in household trash. With textiles, the awareness campaigns may be also challenged by the perspective that used garment exports may destroy apparel manufacturing industries in developing countries; hence, various states implement import stops or high import taxes for used textiles [29].

The awareness campaigns could also involve users in such a way that they separate the electronic or electrically conductive components and the textile substrate and accordingly dispose the different parts into the textile and the electronic waste stream. However, this requires a modular design that enables the separation of the components. The application of the eco-design strategy "design-for-recycling" in the product development process was rated by all except one respondent as an opportunity to improve the recyclability of e-textiles. Furthermore, about 88% agreed that the extension of the scope of the WEEE Directive may lead to an increase in applying design-for-recycling during product development of e-textiles. By extending the scope, e-textiles can be classified under the categories of "Small equipment" and "Small IT and telecommunication equipment" depending on their intended use. Consequently, applying a holistic product planning can facilitate the development of concepts for separate end-of-life processing. Encouragement of research, documentation and evaluation, and the development of best practices may provide access to reliable information and databases in the future. Again, this must be governed by stakeholders from the policy domain, as the recycling sector is unlikely to initiate it. Given the low quantities of e-textiles, there is no apparent reason for the recycling industry to develop solutions.

To drive holistic and efficient recycling of e-textiles, collaboration and sharing of information and best practices among companies in the textile and electronics recycling sectors are essential. Nonetheless, about 88% of the respondents indicated that no cross-sector collaborations existed to date. This may be attributed to the low volume of e-textiles. Hence, industry associations can play a role to facilitate the joint development of processing concepts and assess and suggest suitable processing equipment for e-textiles.

It should be noted that modular product design with the aim of better recyclability is currently still a topic that tends to receive little attention in the field of research and development of e-textile systems. This is also reflected in the low availability of publications on this topic. Modular product design in the field of e-textiles is currently mainly utilized in the context of building kits for rapid and accessible prototype development [30–33]. Fiber-, carbon-nanotube-, and graphene-based electronic and electrically conductive components primarily aim at improving the reliability, comfort, and functionality aspects of e-textiles [17, 34, 35]. Highly integrated electronic and electrically conductive components are difficult to separate from the textile substrate, posing a challenge for sustainable product development. Reparability and maintenance by end users becomes difficult or impossible. Likewise, it is questionable whether product-responsible companies with internal or external repair services can repair highly integrated e-textile components without damaging the textile substrate. Simply replacing defective products should not be considered a sustainable solution, as it would create further waste and resource consumption. Insufficient reparability promotes short product life cycles through premature obsolescence. Nevertheless, it is conceivable that a modular or semi-modular product design would meet with approval from end users and other stakeholders.

## 5. Conclusion and outlook to improve e-textile recycling

The results of the study reveal that e-textiles have so far played a minor role for sorting and recycling companies in Europe: e-textiles are not commonly found products at sorting and recycling companies. Consequently, only about one-third of the businesses already have specialized processing concepts for e-textiles. During sorting, e-textiles are recognized to some extent; however, the technology and machinery of sorting and recycling companies are not designed for the processing of e-textiles. The low waste quantities also lead to a lack of urgency to develop special recycling concepts for e-textiles. Even with a higher market volume, they still will make up a very small percentage of textile and electronic waste streams.

The results of the survey also provide insights into the conditions that must be met to ensure the recycling of e-textiles. Sustainable product development that applies eco-design strategies such as circular design approaches acknowledging end-of life treatment can improve the recyclability of products. It can also help to comply with current and future EU directives and legislation. Especially a modular product design may simplify the separation of e-textiles. This would enable the use of existing processing infrastructures for used textiles and e-waste through collaborations across electronic and textile recycling companies. As currently ease of separation implies also a compromise on longevity of products, e-textiles require novel solutions to integrate electronics and textiles for improved recyclability. Since electronic components interfere with textile recycling if they are not detected, it is advisable to dispose e-textiles at electronic waste collection points.

In order to govern the product design and the e-textile waste treatment, an extension of the scope of the WEEE Directive appears to be fruitful. For example, the definition of small equipment in the WEEE Directive can be adjusted to include e-textiles. The legal frameworks for sustainable, circular product development are also established at EU level by the Ecodesign Directive 2009/125/EC. The scope has so far been limited to energy-related products. Both textiles and products with electrical circuitry and consequently e-textiles are not addressed. Again, expanding the scope of the Ecodesign Directive might facilitate the design of e-textiles for efficient recyclability.

The lack of financially viable business models in recycling compared with low-cost supply of new products impedes the recycling rate. Approaches to integrate the end-of-life treatment in the product costs and distribute the costs partially to the responsibility of the producers may contribute to the development of efficient recycling processes. However, sustainable business models for increasing recycling require definitely further exploration.

## Conflict of interest

The authors declare to have no conflict of interest.

## Thanks

The authors thank the participants of the survey for their contribution.

## Author details

Elisabeth Eppinger[1*], Alina Slomkowski[1], Tanita Behrendt[1], Sigrid Rotzler[2] and Max Marwede[2]

1 University of Applied Sciences for Technology and Economics (HTW), Berlin, Germany

2 Fraunhofer Institute for Reliability and Microintegration (IZM), Berlin, Germany

*Address all correspondence to: elisabeth.eppinger@htw-berlin.de

IntechOpen

## References

[1] Hayward J. E-Textiles 2016-2026: Technologies, Markets, Players. Cambridge, UK: IDTechEx; 2021

[2] Rotzler S, Kallmayer C, Dils C, von Krshiwoblozki M, Bauer U, Schneider-Ramelow M. Improving the washability of smart textiles: Influence of different washing conditions on textile integrated conductor tracks. The Journal of The Textile Institute. 2020;**111**(12):1766-1777. DOI: 10.1080/00405000.2020.1729056

[3] Gonçalves C, Ferreira da Silva A, Gomes J, Simoes R. Wearable E-textile technologies: A review on sensors. Actuators and Control Elements. Inventions. 2018;**3**(1):14. DOI: 10.3390/INVENTIONS3010014

[4] Kirstein T. The future of smart-textiles development: New enabling technologies, commercialization and market trends. In: Kirstein T, editor. Multidisciplinary Know-How for Smart-Textiles Developers. Cambridge: Woodhead Publishing; 2013. pp. 1-15. DOI: 10.1533/9780857093530.1

[5] Korhonen J, Honkasalo A, Seppälä J. Circular economy: The concept and its limitations. Ecological Economics. 2018;**143**(C):37-46. DOI: 10.1016/j.ecolecon.2017.06.041

[6] European Commission. Closing the Loop—An EU Action Plan for the Circular Economy. Brussles; 2015. Available from: https://eur-lex.europa.eu/legal-content/EN/TXT/?uri=CELEX:52015DC0614. [Accessed: July 27, 2022]

[7] Isles, J. Which Country Is Leading the Circular Economy Shift? [Internet]. 2021. Available from: https://ellenmacarthurfoundation.org/articles/which-country-is-leading-the-circular-economy-shift. [Accessed: July 27, 2022]

[8] PIB Dehli. Govt Driving Transition from Linear to Circular Economy [Internet]. 2021. Available from: https://pib.gov.in/PressReleasePage.aspx?PRID=1705772. [Accessed: July 27, 2022]

[9] UNSDG. United Nations Sustainable Development Cooperation Framework for the People's Republic of China 2021-2025. [Internet]. 2021. Available from: https://unsdg.un.org/sites/default/files/2020-11/China-UNSDCF-2021-2025.pdf. [Accessed: July 27, 2022]

[10] Yang K, Isaia B, Brown LJE, Beeby S. E-textiles for healthy ageing. Sensors. 2019;**19**(20):4463. DOI: 10.3390/s19204463

[11] Wilson P, Teverovsky J. New product development for e-textiles: Experiences from the forefront of a new industry. In: Horne L, editor. New Product Development in Textiles. Cambridge: Woodhead Publishing; 2012. pp. 156-174. DOI: 10.1533/9780857095190.2.156

[12] Hayward J. E-Textiles 2021-2031: Technologies, Markets and Players. Cambridge, UK: IDTechEx; 2020

[13] Iftekhar Shuvo I, Decaens J, Lachapelle D, Dolez PI. Smart textiles testing: A roadmap to standardized test methods for safety and quality-control. In: Kumar B, editor. Textiles for Functional Applications. London: IntechOpen; 2021. pp. 141-170. DOI: 10.5772/intechopen.96500

[14] Stoppa M, Chiolerio A. Wearable electronics and smart textiles: A critical

review. Sensors. 2014;**14**(7):11957-11992. DOI: 10.3390/s140711957

[15] Bosowski P, Hoerr M, Mecnika V, Gries T, Jockenhövel S. Design and manufacture of textile-based sensors. In: Dias T, editor. Electronic Textiles. Smart Fabrics and Wearable Technology. Amsterdam: Woodhead Publishing; 2015. pp. 75-107. DOI: 10.1016/B978-0-08-100201-8.00005-9

[16] Simegnaw A, Malengier B, Rotich G, Tadesse M, Van Langenhove L. Review on the integration of microelectronics for E-textile. Materials. 2021;**14**(17):5113. DOI: 10.3390/ma14175113

[17] Seyedin S, Carey T, Arbab A, Eskandarian L, Bohm S, Kim JM, et al. Fibre electronics: Towards scaled-up manufacturing of integrated e-textile systems. Nanoscale. 2021;**13**(30):12818-12847

[18] Rotzler S, von Krshiwoblozki M, Schneider-Ramelow M. Washability of e-textiles: Current testing practices and the need for standardization. Textile Research Journal. 2021;**91**:19-20. DOI: 10.1177/0040517521996727

[19] Eppinger E. Recycling technologies for enabling sustainability transitions of the fashion industry: Status quo and avenues for increasing post-consumer waste recycling. Sustainability: Science, Practice and Policy. 2021;**18**(1):114-128. DOI: 10.1080/15487733.2022.2027122

[20] Wilting, J, van Dujin, H. Clothing Labels: Accurate or Not? [Internet]. 2020. Available from: https://assets.website-files.com/5d26d80e8836af2d12ed1269/5e9feceb7b5b126eb582c1d9_20200420%20-%20Labels%20Check%20-%20report%20EN%20web%20297x210mm.pdf. [Accessed: September 13, 2021]

[21] Stanescu MD. State of the art of post-consumer textile waste upcycling to reach the zero waste milestone. Environmental Science and Pollution Research. 2021;**2021**(28):14253-14270. DOI: 10.1007/s11356-021-12416-9

[22] Anastasio, M. Whatever Happened to Europe's Circular Economy Ambition? [Internet] 2020. Available from: https://meta.eeb.org/2020/11/03/whatever-happened-to-europes-circular-economy/. [Accessed: August 17, 2021]

[23] The European Parliament, The Council of the European Union. Directive 2012/19/EU of the European Parliament and of the Council of 4 July 2012 on Waste Electrical and Electronic Equipment (WEEE). Official Journal of the European Union. 2012. Available from: https://eur-lex.europa.eu/legal-content/EN/TXT/HTML/?uri=CELEX:32012L0019&from=DE. [Accessed: July 27, 2022]

[24] Rucevska I, Nellemann C, Isarin N, Yang W, Liu N, Yu K, et al. Waste Crime—Waste Risks: Gaps in Meeting the Global Waste Challenge. [Internet]. 2015. Available from: https://wedocs.unep.org/bitstream/handle/20.500.11822/9648/Waste_crime_RRA.pdf?se. [Accessed: September 17, 2021]

[25] Wang Z, Zhang B, Guan D. Take responsibility for electronic-waste disposal. Nature News. 2016;**536**:23-25. DOI: 10.1038/536023a

[26] Bogner K, Landrock U. Response Biases in Standardised Surveys. Mannheim: GESIS - Leibniz-Institut für Sozialwissenschaften; 2016. DOI: 10.15465/gesis-sg_en_016

[27] Watson D, Palm D, Brix L, Amstrup M, Syversen F, Nielsen R. Exports of Nordic used textiles: Fate, benefits and impacts. Nordisk

Ministerråd. 2016. p. 160. DOI: 10.6027/TN2016-558

[28] Manshoven, S, Christis, M, Vercalsteren, A, Arnold, M, Nicolau, M, Lafond, E, Fogh Mortensen, L, Coscieme, L. Textiles and the environment in a circular economy. European Topic Centre on Waste and Materials in a Green Economy ETC/WMGE 2019/6. ETC/WMGE Report. [Internet]. 2019. Available from: https://ecodesign-centres.org/wp-content/uploads/2020/03/ETC_report_textiles-and-the-enviroment-in-a-circular-economy.pdf. [Accessed: July 27, 2022]

[29] Brooks A, Simon D. Unravelling the relationships between used-clothing imports and the decline of African clothing industries. Development & Change. 2012;**43**(6):1265-1290. DOI: 10.1111/j.1467-7660.2012.01797.x

[30] Kazemitabaar M, He L, Wang K, Aloimonos C, Cheng T, Froehlich JE. ReWear: Early explorations of a modular wearable construction kit for young children. In: Proceedings of the 2016 CHI Conference Extended Abstracts on Human Factors in Computing Systems (CHI '06); 7 – 12 May 2006; San Jose, USA. 2006. pp. 2072-2080

[31] Woop E, Zahn EF, Flechtner R, Joost G. Demonstrating a modular construction toolkit for interactive textile applications. In: Proceedings of the 11th Nordic Conference on Human-Computer Interaction: Shaping Experiences, Shaping Society (NordiCHI '20); 25-29 October 2020; Tallinn, Estonia. 2020. pp. 1-4

[32] Jones L, Nabil S, Girouard A. Swatch-bits: Prototyping e-textiles with modular swatches. In: Proceedings of the Fourteenth International Conference on Tangible, Embedded, and Embodied Interaction (TEI '20); 9-12 February 2020; Sydney, Australia. 2020. pp. 893-897

[33] Garbacz K, Stagun L, Rotzler S, Semenec M, von Krshiwoblozki M. Modular E-textile toolkit for prototyping and manufacturing. Multidisciplinary Digital Publishing Institute Proceedings. 2021;**68**(1):5. DOI: 10.3390/PROCEEDINGS2021068005

[34] Geim AK. Graphene: Status and prospects. Science. 2009;**324**(5934):1530-1534. DOI: 10.1126/science.1158877

[35] Karim N, Sarker F, Afroj S, Zhang M, Potluri P, Novoselov KS. Sustainable and multifunctional composites of graphene-based natural jute fibers. Advanced Sustainable Systems. 2021;**5**(3):2000228. DOI: 10.1002/adsu.202000228

**Chapter 9**

# Tesla's Circular Economy Strategy to Recycle, Reduce, Reuse, Repurpose and Recover Batteries

*Michael Naor*

## Abstract

The purpose of this research is to explore how Tesla is capable to materialize the circular economy futuristic vision. Specifically, it explains how batteries are recycled, reduced, reused, repurposed, and recovered in order to preserve raw materials and dimmish toxic waste disposal. Tesla extends traveling distance by supercharging stations and repurpose degraded batteries for second-life applications to energize home appliances with its solar panels. Tesla intends to substantially diminish the costs of battery production while increasing range by developing an innovative 4680 tab-less cobalt-free battery. An insight emerging from the study is that the fundamental principles upon the operations management field was established such as the concept of focused factory and Goldratt's theory of constraints stay valid and are applicable towards establishing sustainable manufacturing process at the 21st century.

**Keywords:** tesla, circular economy, recycle, reduce, reuse, repurpose, recover

## 1. Introduction

Circular economy (CE) is defined as an industrial revolution designed to be restorative in nature. It includes utilization of renewable energy sources with an emphasis on five pillars of sustainability: recycle, reduce, reuse, repurpose, and recover in order to preserve raw materials as much as possible. Consequently, it diminishes production of toxic materials and ensures safe disposal which in the case of Millions batteries for electric cars present an environmental hazard.

Major step taken in the 21st century to achieve an ambitious world-class progress towards materializing circular economy vision is by moving towards usage of fully electric vehicles. In order to institutionalize environmental protection by sustainable transportation [1], governments worldwide have been building a global momentum to strictly regulate $CO_2$ and greenhouse gases (GHG) emissions. In the Earth Summit (1992, 154 countries signed a treaty to voluntarily reduce emissions of GHG. One of the summit achievements was the establishment of a GHG pooled inventory shared between countries. Subsequently, over 187 countries have already signed the Kyoto protocol (1997) committing themselves to a reduction of GHG by 5.2% from the benchmark levels of 1990 in order to stabilize the depletion of the atmosphere ozone layer, and combat global warming [2]. More recently, the Copenhagen Summit in

2009 reached an accord that recognizes the necessity to maintain the temperature rise no more than 2 degrees Celsius above agreed threshold.

Tesla manifests a market encroachment attempt to meet the social principles mentioned above [3]. For example, it reached 92% battery cell material recovery in new recycling process of Nickel, Copper, and Cobalt. Furthermore, batteries at end-of-life cycle are reused at homes in conjunction with Tesla solar conglomerate. The companies' executives embarked on a campaign to make the transition to electric vehicles not only in their regional areas but also in the global industrial economy. Specifically, at the end of 2018, Tesla sold its 500,000th car and the next half-million car deliveries will take about 15 months at the current production pace (**Figure 1**).

As of January 2019, Tesla surpassed GM, Ford and BMW to rank the world's 4th most valuable car manufacturer in the stock market. According to Long et al. [4], Tesla is regarded by almost forty percent of customers as role model for future electric car manufacturers because of its elegant innovative attributes, innovation and artificial intelligence technology. It should be mentioned that the hype surrounding its CEO Elon Musk contributes to its stellar image too. In certain states such as California its sales have consistently surpassed over several quarters leading traditional combustion engine brands.

The purpose of this research study is to illustrate how Tesla developed a 21st century manifestation of Ford T mass-production philosophy to sustainable transportation [5]. The similarities between their founders' (Ford and Elon Musk) entrepreneurship skills are stark and brings hope to revive American manufacturing

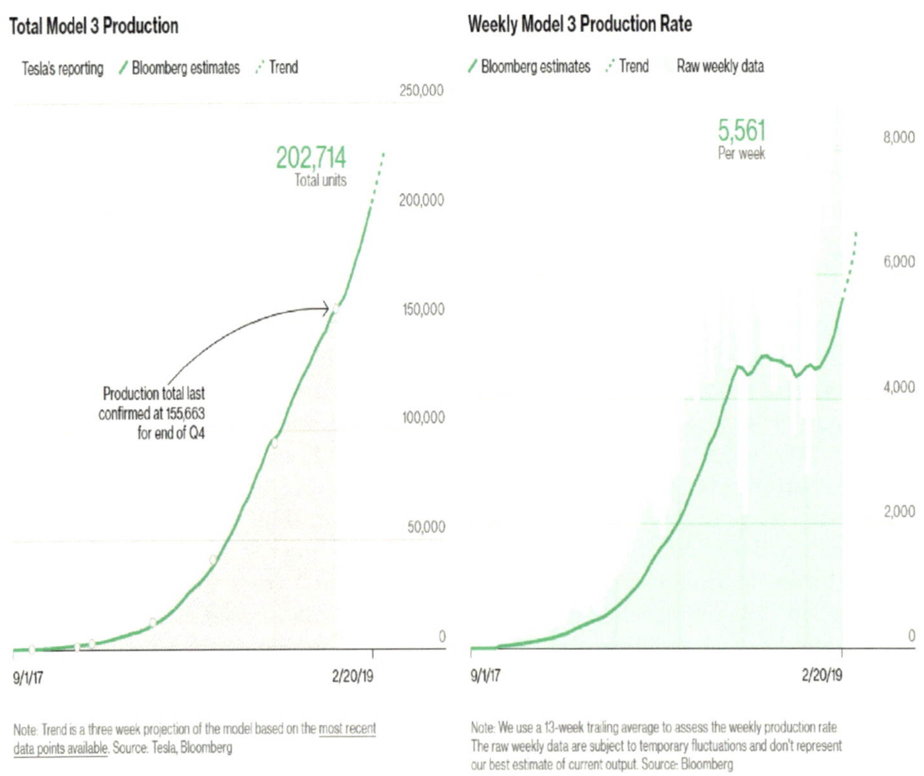


**Figure 1.**
*Tesla production rate based on VIN registration.*

global lead which seemed to be totally under control during last fifty years initially by Japan and more recently by China. Tesla follows an economy-of-scale mindset which is similar to Ford T mass-production line. It gains competitive advantage from capability to recycle, reduce, reuse, repurpose and recover battery materials, all in affordable expenses, which as will be elaborated in this article brings about the competitive manufacturing strategy of Tesla [6]. Tesla possesses vertically integrated supply chain built of Gigafactories producing approximately five thousand cars per week and thousands of batteries/cars [7]. Tesla reduces bullwhip effect phenomena associated with interruption in the supply chain by mining of battery raw materials instead of relying on external suppliers of battery ingredients [8]. Tesla manufacturers far more batteries in terms of kWh than the relevant contestants all together (approximately 15GWh/year, or 0.15TWh).

Prominent scholars such as Harper et al. [9] and Siqi et al. [10] argue that discovery of advanced waste management techniques for batteries is essential for market domination of electric cars. These new technologies include pyrometallurgy usage of high temperature to extract materials. Hydrometallurgy is an innovative technology to recycle metals from ores with low reaction energy consumption. In a similar vein, Biometallurgy introduces a way to extract valuable metals by interaction with microorganisms. The major advantage of this technology is that it entitles lower expenses and render less pollution in comparison with pyrometallurgy and hydrometallurgy.

Substantial amount of saving in materials can be gained by repurposing battery packs towards second-life applications. Leading scholars in the field such as Hua et al. [11] and Yang et al. [12], corroborate that reuse and repurpose processes render less environmental signature in juxtaposition to recycling and recovery of ingredient materials composing batteries due to waste disposal residuals. Thus, Tesla, a solar panels mega-manufacturer, decided to repurpose batteries of old electric vehicles which are no longer capable to efficiently propel a car in order to electrify home appliances as manifestation of circular economy.

## 2. Applying Goldratt's approach to Tesla's needs

The theory of constraints [13] practical approach on generating profit from sales is useful for evaluating bottlenecks of large-scale projects such as the innovation of electric cars. It stands on three pillars: throughput, operational expenses, and inventory [14]. Previous similar electric car mega-projects such as Better Place in Israel were bankrupt because lack of sales rendered by poor marketing. The enormous amount of money required to establish country-wide charging infrastructure demands equally value assets secured in pre-sales format to ensure financial stability, a lesson carefully learned by Tesla from Better Place bankruptcy [15].

Goldratt's principal idea is to locate bottleneck and utilize it efficiently to streamline the process. Traveling range on single full charge has been considered by scholars for over a century the prime bottleneck of electric transportation [16, 17]. Skippon and Garwood [18] empirically substantiate this claim by finding that consumers decision to procure an electric vehicle as a second car is dependent on its ability to travel a distance of 100 miles, and as their first priority vehicle if it has a traveling range exceeding 150 miles.

Tesla philosophy is based on imitating Fort T mass-production line (estimated to reach sales of 1 Million cars by 2021) by establishing advanced gigafactories.

Multi-purpose team is an essential part of this manufacturing paradigm consistent with Deming's quality management philosophy [19].

Thomas and Maine [20], claim that Tesla production mindset did not follow a disruptive innovation route, instead its origin of commercialization success derives from an architectural model based on deploying supercharging stations countrywide which are meant to relieve customer's traveling range anxiety Importantly from a business perspective which has been Tesla strength from the outset because of Elon Musk sensemaking ability, the charging infrastructure is compatible with rival's electric car design too, rendering an additional source of income from offering charging services to rivals which contributes to its image as green company with altruism motive, instead of greedy hidden agenda driven by ulterior motive to become a monopoly.

## 3. Methodology: case study description of mega-project

Shenhar and Holzmann [21] highlight that there is void in the literature in the subject of how to manage complex mega-projects. This is especially evident in pioneer megaprojects which require high degree of adaptation due to lack of experience. Tesla can be classified among this cluster of unprecedented projects both in its technological novelty and magnitude. As such, the method of case study is appropriate [22]. A longitudinal series of rigor interviews over a decade with Tesla employees as well as similar mega projects such as EV-1 in California and Better Place in Israel constituted the empirical approach in this case. It was validated with secondary data sources to juxtapose sources and verify information. Face-to-face interviews were conducted with Tesla's marketing, procurement, and technical engineering in two headquarter centers located in McLean, Virginia and Washington, DC.

## 4. Interpretation of results

In 2003, several engineers at Silicon Valley embarked on a journey to advance the world's movement to green mobility [23]. To make this century-long dream come true, Tesla Motors was founded. As will be seen, Tesla always went bigger than industrial benchmark, appropriately setting its production quantities to yield sales of 1 million cars by 2021.

In the first step, on 2008, Tesla introduced the Roadster, a car with capability to drive 245 miles on single charge. Afterwards, Tesla introduced the Model S, with a traveling distance of 265 miles, which acclaimed Motor Trends' 2013 car of the year prize. Next, it began manufacturing the Model X, a crossover type car with an additional third seating row. Tesla is currently building $5 billion worth battery factory that is estimated to produce more lithium-ion batteries in 2020 than all of the contestants' yield.

Tesla owners are able to charge their batteries to 50% level in a short timeframe of 20 minutes. Tesla expanded rapidly in 2014, starting with charging stations in Norway, afterwards encroached 12 additional countries, and plans to enter into the market of almost every country in Europe [24]. In effort to extend international sales, Tesla began marketing its electric vehicles to the Chinese niche on August 2013. Although China's market is larger than Europe, the regulatory environment in Europe makes it a favorable destination for electric cars in the next decade because European Parliament Transport Committee approved a resolution in November 2013 making it compulsory that EU country members need to install network of at least one charging station per 100 km.

Tesla Motors portfolio is diversified to include both vehicle and battery components. The batteries manufactured by Tesla are compatible with other car brands extending their market worldwide. For example, Toyota and Mercedes utilize Tesla's battery in the Rav4 and Mercedes B-Class.

In contrast to Better Place infrastructure which was based on the notion of swapping depleted batteries (**Figure 2**) of cars in short time period of 5 minutes but was limited to single car manufacturer (Renault), Tesla supercharging stations are capable to rapidly charge depleted battery in forty minutes. The super-charging stations are compatible with various car manufacturers rendering the infrastructure a servicizing source of income for Tesla. As of April 2019, Tesla had a network of 12,000 supercharging stations across North America, Europe and Asia. According to Tesla, the funding needed to establish a supercharging station is $150,000 without solar panels, and $300,000 to construct a solar powered facility. A single station can charge multiple cars simultaneously (usually 2–4 spots are available). The decision where to build a super-charging station is determined based on actual electric car sales and strategically curated after rigor post-sales survey of consumers' driving patterns.

On March 2019, Tesla debuted an innovative V3 supercharger architecture that is capable to reduce the charging time on average by approximately 50%. Furthermore, utilizing the V3 supercharger, the Model 3 car extends its traveling range by an additional 75 miles on a quick charge that consumes short duration of five minutes. To avoid over-usage of supercharging services which degrades the battery's lifespan, the V3 supercharger can charge a battery up to 80% instead of its maximum capacity in 45 minutes because subject matter experts in university investigation found that charging a battery to its full extent had negative impact on its lifespan. Interestingly, Tesla is debuting a technology titled, on-route battery warmup, which heats the battery to an optimal temperature on the way to the supercharger station. The warmup technology diminishes the average charge time by an additional 25%. Overall, V3 supercharger network permits Tesla to double the amount of vehicles it servicizes in order to meet the needs of its exponentially increasing fleet.

Tesla portfolio has major differences from past mega-projects attempting to electrify transportation. The innovator's dilemma is whether to focus on becoming solely

**Figure 2.**
*Renault Fluence swapping station better place.*

an electric car manufacturer as Tesla have chosen or to diversify portfolio with blend of fuel combustion cars too such as done by other mainstream manufacturers such as Toyota [25]. Tesla pursues Skinner's [26] seminal model called, focused factory, by creating array of Gigafactories which builds fully electric vehicles (not including hybrid or combustion engine type of cars inside it). Its Model X and S target different household incomes. The Model 3 represents an affordable family sedan with distance of 320 miles per charge, Model S large sedan with distance of 400 miles and Model X is a large SUV traveling 300 miles. Model Y is a mid-size SUV with 315 miles traveling distance. In future, the second-generation coupe Roadster is going to be designed in order to travel 650 miles per charge.

Tesla marketing is using social media word of mouth as its main networking tool to reach audience. Surprisingly, this low budget method achieves record high pre-orders which paces the production line capacity.

Trying to stay head of the curve, Tesla vehicles come equipped with autonomous capability based on variety of technologies such as radar, sonar, acoustic sensors and a network of cameras to identify pedestrians, cars and other potential obstacles on the route. The United States Department of Transportation categorizes Tesla as Stage 2 level self-drive capability, meaning that a driver must be seated behind the wheel at all times. Tesla self-driving parallel parking capability has been appealing for wide segments of population. The capability for stage 3 autonomous driving without driver behind the wheel is built-in by Tesla vehicles too, but its pending approval of regulatory agencies.

To meet weekly pre-sale, Tesla emulates Ford T mass-production mindset by building Gigafactories. Historically, the Ford T used the stationary construction methods available in the 20th century, assembly by hand, to manufacture small batches. The original Ford Piquette Avenue Plant could not meet demand for the Model T because 11 cars were built there during the first month of production. Consequently, in 1910, in order to create mass-production line, Henry Ford moved the factory to the new Highland Park facility. The Model T production line shifted into an innovative modular format where Ford's cars were constructed rapidly, diminishing production time from 12.5 hours beforehand to 93 minutes by 1914. It was accomplished by conveyor belts, a technology which standardized the process. This allowed Ford to decrease the cost of cars by gaining economies of scale [27, 28]. Fredrick Taylor rendered consulting services to diminish the assembly line into 84 discrete steps (an unprecedented accomplishment last century). Subsequently, Ford built machines that could stamp-out large car components automatically for engine and transmission.

After the Model T reached a remarkable threshold of 10 million vehicles, it accounted for 50% of all cars globally. Similar to Tesla marketing method, Ford brand was notably famous among consumers so it did not require extra advertisement. By 1925, Ford plant reached a production milestone of 15 million vehicles, with a manufacturing pace of 10,000 cars per day (2 million annually).

Following Ford's footsteps, Tesla established hubs (Gigafactories) around the world to mass-produce electric cars. These green plants are energized by solar power. Gigafactory 1 located at Nevada is a large size, 10 million square feet plant, which is forecasted to yield 500,000 batteries for the Models S, X and 3 cars per year. The facility also produces Tesla Powerwall, Powerpack, and Megapack devices. It is expected to build the Tesla Semitruck in future. Battery production at Gigafactory 1 passed an annualized pace of about 20 GWh (3.5 Million cells per day), ranking it the highest-volume battery plant globally. Tesla has been innovating Tab-less cobalt-free lithium battery cell measuring 46 by 80 millimeters (hence called 4680), utilizing nickel-manganese structure, which extends by 16% the car range and multiply six times the amount of energy, while decreasing expenses to produce the cell by 14% compared to existing cells which are powering the model 3 and Y. Its anode uses raw metallurgical silicon which does not crack. The new cylindrical architecture shape batteries will utilize Maxwell's dry electrode technology which is lowers cost and is purer than lithium-ion batteries. The rolled-up copper material cuts the distance for electrons to travel rendering a decrease in internal resistance and heat dissipation. The state of Nevada, where Tesla Gigafactory is located has plenty deposits of lithium embedded in clay which Tesla plans to mine. Overall, Tesla plans to diminish the price per kilowatt hour (kWh) of its cells by 56%. After the Model 3 production had reached the pace of 5000 cars a week, the manufacturing quota of battery cells in the Gigafactory had reached 3.5 million cells per day. As of 2019, the plant employs about 8000 workers.

Gigafactory 2, Tesla's plant for Model 3 cars in Fremont, California, was in the beginning constructed on 5.3 million square feet but latter expanded after the City of Fremont's granted permission for Tesla, in 2016, to double the plant's area to 10 million square feet, employing over 10,000 people in the Fermont's municipality. It is connected to rail network which transports batteries and parts between Tesla Gigafactories. The production line uses more than 160 robots, including 10 of the biggest robots globally. The assembly process takes between three to five days. The battery pack which is put inside car weighs approximately 1200 pounds. Tesla built a mega casting machine in the Fremont factory in order to produce large car parts in a single piece. The one-of-a-kind machine called, Giga Press, made of aluminum die casting, produced by Idra Group in Italy, has a clamping force of about 60,000 Kilonewtons. It reduced casting from 70 parts to four parts, with future goal to produce most of the Model Y frame in one piece.

Gigafactory 3, is located in Shanghai, China. The plant was built in a minimal time period of one year. Tesla targets this plant to produce a quota record of 250,000 cars per year (starting with a weekly production quota of 3000 units). Tesla introduced the Model Y crossover for the Chinese market, postulating based on the assumption that sales for this car model are going to exceed that of Tesla's other models altogether. The president of Tesla's location in China, proclaimed that it is designing a cheaper Tesla for the low-income customer niche which is expected to be priced at an affordable $25,000. It is called Model 2 and will be a hatchback.

Gigafactory 4, is at Austin, Texas. This plant is going to manufacture the cybertruck (**Figure 3**) which has over 500,000 preorders. Tesla is developing three models of cybertruck: single, dual and tri motor) to conquer the vast market of electric pickup trucks which is very popular in North America rural landscape. The cybertruck is built with an exterior shell called exoskeleton for extra strength, and equipped with a passenger protection armor glass.

Gigafactory 5, is taking ground at Berlin, Germany. The plant is going to produce Model Y (500,000 car annually) and associated battery packs. Tesla ordered eight new casting equipment devices that are planned to be assembled in the Berlin Gigafactory which encroaches into the European market.

There is a plan in blueprint to establish Gigafactory 6, at India. Specifically, on January, 2021, Tesla stepped into the Indian market by officiating a subsidiary, Tesla India Motors and Energy Private Limited. Tesla's Indian branch is considered to be built in the Karnataka state which encompasses research and development infrastructure hubs. Since India has demand for small size and cheaper price vehicles, the low-cost Model 2 is primary designated for production in this Gigafactory. Importantly, India's interest in the worldwide growing trend of sustainability fits with Tesla solar panels branch.

**Figure 3.**
*Tesla Cybertruck.*

## 5. Discussion of tesla business model limitations and future challenges

Tesla exponential sale's grow is notable, but it faces numerous impediments in the near future. First, the battery production is composed of three components: cell manufacturing, module manufacturing, and pack assembly. Tesla manufactures battery modules and packs at both its Gigafactory in Nevada, and at its vehicle assembly plant in Fremont, California. Tesla's battery packs for the Model 3 utilize cells from the Gigafactory, while cells for the Model S and Model X are still manufactured by Panasonic in Japan, which should be brought back home for manufacturing instead of outsourced because pack assembly needs to be near the vehicle assembly location for logistic reasons to reduce cost of transporting battery packs, which are larger and heavier than cells or modules. Thus, Tesla is aiming to achieve cell manufacturing capacity of 100GWh per year by 2022 and 3TWh per year by 2030. Through usage of nickel and lithium resources available within North America, and manufacturing cells in-house at Nevada, Tesla may substantially lower the cost of lithium production by 33% and decrease the miles traveled for the battery packs assembly by 80%. Following this mindset, Tesla vertical supply chain should extend into the mining industry coupled with investment in developing frontier material science in collaboration with prime universities which are leading the field. Tesla plans to reduce by 50% the price of its batteries in an effort to sell its flagship Model 2 car in a low-cost price of $25,000. Towards this goal, Tesla signed a contract with North Carolina-focused mining group Piedmont Lithium to purchase five years of their yield starting in 2022 and Tesla is also looking at creative ways to mine lithium from clay deposits in Nevada. These actions are going to render a fraction of Tesla lithium consumption, while the rest can be procured from Chile, Australia, China, etc. Tesla needs 25,000 tons of lithium per year to produce 35 GWh. It is a source of discrepancy because a future Tera-factory can consume up to 800,000 tons of lithium per year. Currently, there is no shortage of lithium worldwide. Which has decreased lithium cost (at 2020's global lithium production output is at 300,000 tons per year), but this situation is going to dramatically change due to disruption in worldwide supply chain rendered by COVID-19 crisis starting 2021 until at least 2024.

Also, the Tesla's Model 3, which is relatively affordable electric car, suffered from production delays. It has recently met the monthly production quota needed to satisfy customers' backlog of reservations and the demand forecasted. Although, Tesla has fulfilled a goal of producing 5000 Model 3 s each week and delivering over 500,000 cars, its ability to meet increasingly high pre-orders rendered by spiking oil prices is doubtful [29]. Consequently, Tesla ended its customer referral program, cut its workforce by 7% and reduced the prices of the Models 3, S and X by $2000 because its sales in the fourth quarter of 2018 fell behind estimates. Tesla has yet to fulfill its major promise to market its flagship Model 3 car at a price of $35,000 and has reduced its production of the Models S and X that are in less demand to 2000 cars per week.

Another source of concern is the quota limit on tax credits [30]. Electric car buyers are granted a $7500 break on their federal taxes. These perks are intended to put more green vehicles on the road. However, each electric car brand has a quota limit: 200,000 units. After passing this threshold, the incentives halt. As of 2019, more than 370,000 consumers have secured deposits for a Model 3 but not all of them will be able to receive the tax exemption, rendering risk of abolishing pre-orders.

A bigger source of concern for Tesla future prospects is that it falls behind rivals in offering car-sharing services. For instance, the Waymo originally developed by Google which pioneered the ride-sharing market, Zoox a self-driving car company that Amazon bought, and AutoX a Chinese self-driving startup funded by Alibaba. BMW group and Daimler AG are pooling resources to invest 1 Billion Dollars in total to develop ride-hailing services. In the future, Tesla plans to further expand its supply chain network by creating a ride-hailing platform titled Robotaxi. According to ARK invest report, this market has the potential to generate 1 Trillion Dollars in annual operating earnings by 2030 [31]. Essentially, the Robotaxi service aims to repurpose the car by allowing Tesla customers whenever not driving their vehicle, to use it as an autonomous taxi.

Finally, one should not ignore that the electric grid can become overloaded on peak-hours of consumption. It requires collaboration between all electric car manufacturing and their affiliated charging stations for development of a smart grid.

## 6. Conclusion

Despite concerns mentioned above, a report by UBS [32] asserts Tesla's future is promising. The bank highlights Tesla's record-high order backlog, growing margins, and an advantage in critical supply chains components. The analysts of the bank corroborated their arguments based on Tesla's two new plants in Germany and Texas, which are ramping up manufacturing and should roughly double the company's production capacity over time. They forecast the new factories with Tesla's pricing is going to keep the company's automotive margins above 30% for quarters to come. Tesla postulates sales grow by approximately 50% on average year over year, a goal UBS confirm it will meet in 2022. Despite lost sales due to the COVID-19 lockdown in Shanghai which restricted employees from attending work for a while, Tesla should be capable to manufacture 1.4 million vehicles on 2022 and reach 2.9 million by 2025. Finally, Tesla's capability to retain competitive advantage stems from software's scalability, which can gain substantial revenue beyond 2025 because Tesla is yet again head of the curve by investing in lidar and machine learning technologies for self-driving cars.

## Author details

Michael Naor
Department of Operations Research, Hebrew University, School of Business Administration, Jerusalem, Israel

*Address all correspondence to: michael.naor@mail.huji.ac.il

## References

[1] Wolfson A, Tavor D, Mark S, Schermann M, Krcmar H. Better place: A case study of the reciprocal relations between sustainability and service. Service Science. 2011;**3**(2):172-181

[2] Naor M, Bernardes ES, Druehl C, Shiftan Y. Overcoming barriers to adoption of environmentally-friendly innovations through design and strategy: Learning from failure of an electric vehicle infrastructure. International Journal of Operations & Production Management. 2015;**35**(1):26-59

[3] Stringham EP, Miller JK, Clark JR. Overcoming barriers to entry in an established industry: Tesla motors. California Management Review. 2015;**57**(4):85-103

[4] Long Z, Axsen J, Miller I, Kormos C. What does tesla mean to car buyers? Exploring the role of automotive brand in perceptions of battery electric vehicles. Transportation Research Part A: Policy and Practice. 2019;**129**:185-204

[5] Hardman S, Shiu E, Steinberger-Wilckens R. Changing the fate of fuel cell vehicles: Can lessons be learnt from tesla motors? International Journal of Hydrogen Energy. 2015;**40**(4):1625-1638

[6] Crabtree G. The coming electric vehicle transformation. Science. 2019;**366**(6464):422-424

[7] Mangram ME. The globalization of tesla motors: A strategic marketing plan analysis. Journal of Strategic Marketing. 2012;**20**(4):289-312

[8] Mann MK, Mayyas AT, Steward DM. Supply-Chain Analysis of Li-Ion Battery Material and Impact of Recycling (No. NREL/PO-6A20-71724). Golden, CO (United States): National Renewable Energy Lab (NREL); 2019

[9] Harper G, Sommerville R, Kendrick E, Driscoll L, Slater P, Stolkin R, et al. Recycling lithium-ion batteries from electric vehicles. Nature. 2019;**575**(7781):75-86

[10] Siqi Z, Guangming L, Wenzhi H, Juwen H, Haochen Z. Recovery methods and regulation status of waste lithium-ion batteries in China: A mini review. Waste Management & Research. 2019;**37**(11):1142-1152

[11] Hua Y, Liu X, Zhou S, Huang Y, Ling H, Yang S. Toward sustainable reuse of retired Lithium-ion batteries from electric vehicles. Resources, Conservation and Recycling. 2020;**168**:105249

[12] Yang J, Gu F, Guo J. Environmental feasibility of secondary use of electric vehicle lithium-ion batteries in communication base stations. Resources, Conservation and Recycling. 2020;**156**:104713

[13] Goldratt EM, Cox J. The Goal. revised ed. Great Barrington, MA: The Northern River Press Publishing Corporation; 1986

[14] Naor M, Bernardes SE, Coman A. Theory of constraints: Is it a theory and a good one? International Journal of Production Research. 2012;**51**(2):542-554

[15] Ito N, Takeuchi K, Managi S. Willingness-to-pay for infrastructure investments for alternative fuel vehicles. Transportation Research Part D: Transport and Environment. 2013;**18**:1-8

[16] Axsen J, Kurani KS, Burke A. Are batteries ready for plug-in hybrid buyers? Transport Policy. 2010;**17**(3):173-182

[17] Kirsch DA. The Electric Vehicle and the Burden of History. Piscataway, NJ, United States: Rutgers University Press; 2000

[18] Skippon S, Garwood M. Responses to battery electric vehicles: UK consumer attitudes and attributions of symbolic meaning following direct experience to reduce psychological distance. Transportation Research Part D: Transport and Environment. 2011;**16**(7):525-531

[19] Anderson JC, Rungtusanatham M, Schroeder RG. A theory of quality management underlying the Deming management method. Academy of Management Review. 1994;**19**(3):472-509

[20] Thomas VJ, Maine E. Market entry strategies for electric vehicle start-ups in the automotive industry–lessons from tesla motors. Journal of Cleaner Production. 2019;**235**:653-663

[21] Shenhar A, Holzmann V. The three secrets of megaproject success: Clear strategic vision, total alignment, and adapting to complexity. Project Management Journal. 2017;**48**(6):29-46

[22] Shenhar AJ, Holzmann V, Melamed B, Zhao Y. The challenge of innovation in highly complex projects: What can we learn from Boeing's Dreamliner experience? Project Management Journal. 2016;**47**(2):62-78

[23] Perkins G, Murmann JP. What does the success of tesla mean for the future dynamics in the global automobile sector? Management and Organization Review. 2018;**14**(3):471-480

[24] Chen Y, Perez Y. Business model design: Lessons learned from tesla motors. In: Towards a Sustainable Economy. Cham: Springer; 2018. pp. 53-69

[25] Christensen CM. The Innovator's Dilemma: When New Technologies Cause Great Firms to Fail. Boston, MA, USA: Harvard Business Review Press; 2013

[26] Skinner W. The Focused Factory. Boston, MA, USA: Harvard Business Review; 1974. pp. 114-121

[27] Alizon F, Shooter SB, Simpson TW. Henry ford and the model T: Lessons for product platforming and mass customization. Design Studies. 2009;**30**(5):588-605

[28] Brooke L. Ford Model T: The Car That Put the World on Wheels. Minneapolis, MN, USA: Motorbooks International; 2008

[29] Bloomberg. Tesla model 3 tracker. 2019. Available from: https://www.bloomberg.com/graphics/tesla-model-3-vin-tracker/?terminal=true

[30] Shiftan Y, Albert G, Keinan T. The impact of company-car taxation policy on travel behavior. Transport Policy. 2012;**19**(1):139-146

[31] Big ideas report. ARK invest. 2021. Available from: https://ark-invest.com/big-ideas-2021/?utm_campaign=Big%20Ideas%202021&utm_medium=email&_hsmi=108239947&_hsenc=p2ANqtz-_DosaDBniWZ3ZtBhhpnLmcjIPKY5kt20hxNGb710eUTGnPbk3MwSwEs3Ys9-VpCRSq7nSSPQJQSjbQ6Tp2MEPvF8de-g&utm_content=108239947&utm_source=hs_email

[32] Business Insider. Tesla's Future 'Brighter Than Ever,' UBS Analysts Say. 2022. Available from: businessinsider.com